Homeowner's Guide to Plumbing

Robert Scharff

Ideals Publishing Corp.
Milwaukee, Wisconsin

Table of Contents

ISBN 0-8249-6106-4

Copyright © 1981 by Ideals Publishing Corporation

All rights reserved. This book or parts thereof may not be reproduced in any form without permission of the copyright owners. Printed and bound in the United States of America. Published simultaneously in Canada.

Published by Ideals Publishing Corporation
11315 Watertown Plank Road
Milwaukee, Wisconsin 53226

Editor, David Schansberg

Drawings by Donald W. Holohan and Robert Mull

Cover designed by David Schansberg. Materials courtesy of Elm Grove Ace Hardware.

Cover photo by Jerry Koser

⌂ SUCCESSFUL
HOME IMPROVEMENT SERIES

Bathroom Planning and Remodeling
Kitchen Planning and Remodeling
Space Saving Shelves and Built-ins
Finishing Off Additional Rooms
Finding and Fixing the Older Home
Money Saving Home Repair Guide
Homeowner's Guide to Tools
Homeowner's Guide to Electrical Wiring
Homeowner's Guide to Plumbing
Homeowner's Guide to Roofing and Siding
Homeowner's Guide to Fireplaces
Home Plans for the '80s
How to Build Your Own Home

How a Home Plumbing System Works

Home plumbing is neither mysterious nor overly complicated. On the contrary, it is relatively easy to understand, especially when you become familiar with the tools and the wide assortment of products available. Today, with standardized fittings and tools available, almost anyone can acquaint himself or herself with the general requirements and do a creditable job of plumbing. But, before proceeding with any plumbing job, be sure to check your local plumbing or sanitary code. There are still a few cities in which all work on a residential plumbing job must be done by licensed plumbers. In some localities, the code states that all soil and drain pipes must be installed by professionals, while a few communities permit anyone to do the work provided it is inspected and approved by licensed plumbers upon completion.

In any case, it is a good idea to obtain a copy of the local plumbing code and study it carefully. The rules that appear in it usually follow the generally accepted practice of good plumbing. Remember that good plumbing does not depend as much on who does the job as it does on how the job is done. The comfort and health of your family depend on good workmanship and common sense when working on your home's plumbing system.

The total plumbing system includes all pipes, fixtures, and fittings used to convey water into and out of the home. It can be divided into three basic areas: (1) the water-supply system; (2) the drainage system; and (3) the fixtures. The object of it all, of course, is to make water available where needed in the home, and to get rid of the water, plus wastes, after it has served its purpose.

The Water-Supply System

In every plumbing system, there must be a source of water and the pipes to carry it to the fixtures. This water-supply system must be adequate to:

- Assure you of pure water for drinking.
- Supply a sufficient quantity of water at any outlet in the system, at correct operating pressure.
- Furnish you with hot or cold water, as required.

In most incorporated areas, the water source is a public or privately operated water "works" from which purified water is distributed through mains to which each user can be connected by arrangement with the proper authorities. However, if such a source is not available, you can install a private source of your own. In the latter case, the most efficient source in rural areas is the drilled well. In most cases, this is a job for a professional well digger. Wells are expensive and require special equipment for drilling. Local experts will be able to provide the answers to your questions concerning locating the well and its necessary equipment.

Cold-Water System The main supply line coming into a house carries cold water. Normally, it can be turned off near its point of entry, usually just beyond the water meter, if there is one. From this main supply, branches lead to the various fixtures. The supply mains should be graded to one low point in the basement or crawl-space so that a drain cock will permit complete drainage of the entire supply system. Any portion of the piping which cannot be so drained must be equipped with a separate drain cock. As a rule, a pitch of ¼ inch to each foot of pipe is sufficient to permit proper drainage.

Hot-Water System Hot water is obtained by routing cold water through a water heater. This heater may be part of the central heating plant or a separate unit. When part of a central system, a separate hot-water storage tank is generally provided to hold the heated water. On the other hand, when a separate heater is used, the water is stored within the unit.

A separate heater unit may be electric, oil, or gas fired, but all are automatically controlled by a preset thermostat. Each style of heater comes in a wide variety of sizes. All automatic heaters have the necessary internal piping already installed, and the only connections required are the hot- and cold-water and fuel lines. Oil or gas-fired water heaters also require flues to vent the products of combustion.

A new hot-water heater might be necessary when you add new plumbing to your present system. Even if the present water heater is functioning properly, there may not be sufficient hot water available for your family. Actually, the size of the hot-water storage tank needed in the house depends upon the number of persons in the family, the volume of hot water that may be needed during peak use periods and the "recovery rate" of the heating unit. A good rule to follow when estimating the capacity of the tank required is 10 gallons per hour for each member of the family. For a family of four, for example, the hot-water demand will be 40 gallons of hot water per hour. This does not mean that the system will operate continuously at that capacity, but it must be capable of producing that amount of hot water to keep up with normal usage. If any unusual demands are anticipated, a larger capacity should be provided.

The recovery rate of water heaters varies with the type and capacity of the heating element. In standard conventional models, oil and gas heaters usually have higher recovery rates than electric heaters of similar size. For this reason, you would want a slightly larger capacity of electric heater than for either oil or gas. Another option is the quick recovery type of electric heater.

Temperature and pressure (T&P) relief valves are on all hot-water heaters and hot-water storage tanks. Their function is to relieve pressure in the tank and water pipes should any other piece of control equipment in the system fail and the water temperature reach a point high enough to cause a dangerous pressure that would rupture the tank and pipes. Another important device on the heater is the drain cock or valve. Located at the bottom of the storage tank, it allows for the draining of the tank. A shutoff is also located on the cold-water intake pipe.

Once the hot water leaves the heater, it runs through a system of pipes similar to but separate from the cold-water system. It passes through hot-water

mains, branch lines, and fixture supply lines, then ends at the fixtures. The hot-water valve or faucet is always placed on the left side of a fixture as you face it.

If a fixture requires only cold water (such as a toilet), the hot-water main passes that fixture untapped. Often, the hot-water supply is divided into two mains, one to supply water at 140 to 160 degrees for dishwashers and washing machines, and one to supply water at 120 degrees for safe use at sinks, showers, and bathtubs. The most efficient system of hot-water piping is a continuous loop in which the hot water constantly recirculates through the mains and back to the heater. Of course, the recirculation system does not have anything to do with recirculating used hot water; the water-supply system and the drainage system are entirely separate.

Drainage System

Drainage (strictly controlled by code in most localities) is the complete and final disposal of the waste water, the sewage it contains, and the gas which the sewage decomposition produces. A drainage system, therefore, consists of: (1) the pipes that carry sewage away from the fixtures; (2) the place where the sewage is deposited; and (3) the system of vents which allow sewer gas to escape. You may empty sewage into a city sewer, into a properly constructed septic tank, or (in a few cases) into a cesspool.

In the house, the waste lines are concealed in the walls and under floors. The vertical lines are called stacks, and the horizontal lines are called branches. The flow of waste water starts at the fixture trap, the device that stops sewer gases from entering the house. It flows through the fixture branches to the soil stack. It continues through the house drain and the house sewer and finally reaches the city sewer or, in a private system, a septic tank. Waste stacks carry only water waste. The lines taking the waste from the toilet are called the soil lines and soil stacks. Because of the solid waste materials, soil lines and stacks are the largest in the system. Each time the soil lines are used, they are flushed. Also needed in a drainage system are the vents for the circulation of air.

To be safe, your drainage system has to meet five basic requirements:

1. All pipes in this system must be pitched (slanted) down toward the main disposal so that the weight of the waste will cause it to flow toward the main disposal system and away from the house. Because of gravity flow, the waste lines must be larger than the water-supply lines, in which there is pressure. Fittings are designed with less than 90 degree angles to help maintain the slope.

2. Pipes must be fitted and sealed so that sewer gases cannot leak out.

3. The system must contain vents to carry off the sewer gases to where they can do no harm. Vents

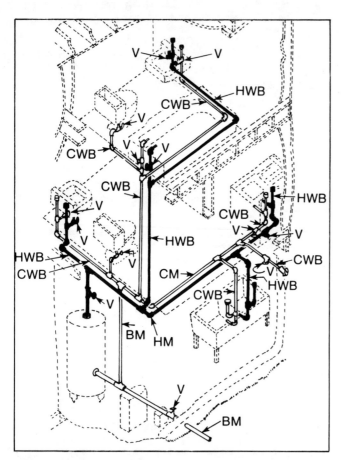

A typical water-supply system: BM = building main; CM = cold water main; HM = hot water main; CWB = cold-water branch; HWB = hot-water branch; V = valve.

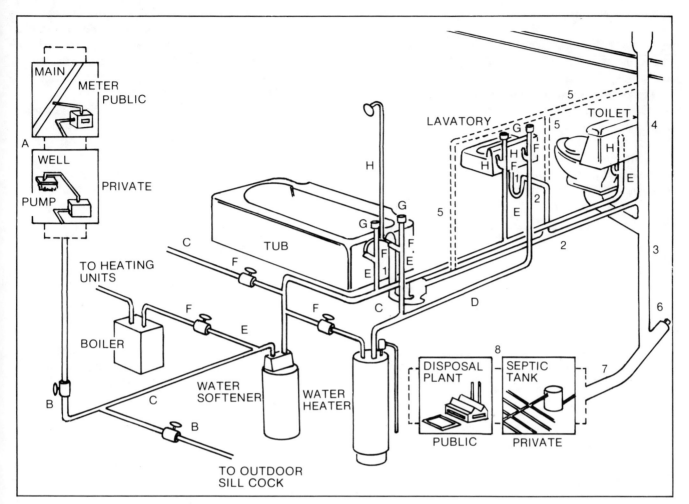

Major parts of the plumbing system: Supply system parts: (A) Water source, public or private. For our purpose it includes all piping up to the building. (B) Drain cock or shutoff valve. One is required at the low point of the system. (C) Cold-water main line (any line serving two or more fixtures). (D) Hot-water main line (any line serving two or more fixtures). (E) Branch line to fixture (any line, for cold or hot water, that serves one fixture only). (F) Shutoff valve, recommended for use in all branch lines and in main lines wherever a cutoff might be required. (G) Air chamber, recommended for any branch line terminating in a faucet. It helps eliminate chatter. (H) Fixture supply line. This is the portion of a branch line (above) that is installed when the fixture itself is installed and is adapted to its special requirements. Drainage system parts: (1) Fixture drain, the portion of a branch drain (below) adapted to the requirements of the particular fixture. Each drain must incorporate a trap (unless a trap is built into the fixture) that will hold water and

seal the drain line against the escape of gases into the house. (2) Branch drain, a line between a fixture and a soil stack. (3) Soil stack, a vertical pipe that collects from the branch drain or drains. Every installation must have one main stack, that is, a stack built of 3- or 4-inch pipe (depending on the building code), extending all the way through the roof. There may be a secondary stack or stacks, built of smaller pipe (usually 2-inch), either throughout or in the vent portion only. (4) Vent, upper portion of a soil stack, through which gases escape to the outside and air enters the stack. (5) Re-vent, a by-pass for air between a branch drain and the vent portion of a stack. It is required by some codes. (6) Cleanout. One should be placed at every point where access may be needed to clear an obstruction (always at the foot of each stack). (7) Building drain. It receives waste from the stack or stacks and carries it to the final disposal. (8) Final disposal, either the public sewer or a private septic tank.

also help to equalize the air pressure in the drainage system. The major vent is the upper portion of the soil stack which extends through the roof.

4. Each fixture that has a drain should be provided with a suitable water trap, so that water standing in the trap will seal the drain pipe and prevent the backflow of sewer gas into the house. The trap for the toilet (water closet) is built into it.

5. Re-vents should be provided wherever there is danger of siphoning the water from a fixture trap or where specified by local codes.

As we have seen, there is an important difference between the water-supply system and the drainage system. In the water-supply system, water flows under pressure; in the drainage system, gravity causes the flow. Therefore, keep these points in mind: the supply system is continuous and closed; the drains are always pitched downward from the fixture to which they are attached and must always be vented. The design of household plumbing, and especially placement of fixtures, often depends on the drainage system, because the pipes are large and hard to install or modify.

Traps and Venting Gases develop in sewers and septic tanks and flow back through the drainage piping system. To prevent these gases from backing up through open fixture drains or overflows and escaping into the house, a trap is required at each fixture. The trap should be the same size as the drain pipe and as close as possible to the fixture outlet. The water seal in the trap should be at least 2 inches, but not more than

4 inches. Water closets usually have built-in traps and no additional one is required. Never double-trap a fixture. Drum traps are commonly used in bathtub drain lines. A trap should be 3 or 4 inches in diameter, and the bottom or top should be removable to permit cleaning of the trap and drain pipe.

Traps prevent sewer gases from entering the house, but sewer gases that are confined can develop pressure and bubble through the water seal in fixture traps. Therefore, at least one vent must be provided through which these gases can escape to the outside air and thus prevent any buildup of pressure or vacuum on the trap seal. This is usually the top half of the main soil stack.

The soil stack should always be vented to the outside, above the roof, and undiminished in size. Toilet, sink, bathtub, shower, and all other drains may be drained directly into the main soil stack without a separate vent if they are within several feet of it. This is called "wet venting." Fixtures that are farther away need a separate vent system running from the fixture to the roof or into the main soil stack vent above the highest fixture connection. The vent pipes that connect into the main vent are called re-vents. Horizontal re-vent runs are always pitched slightly upward to lead the sewer gases up and out. Plumbing codes specify the venting required.

Floor Drains Floor drains are required in shower stalls, of course, but they are often installed in laundry rooms, basements, and utility rooms. Floor drains should be trapped. If the building drain is laid under the floor, it must be at a sufficient depth to permit installation of the trap. Floor drains are usually set close enough to the building drain to eliminate separate venting.

A floor drain should be flush with the floor, and the floor should slope toward the drain from all directions. The grating of the drain should be removable so the drain can be cleaned.

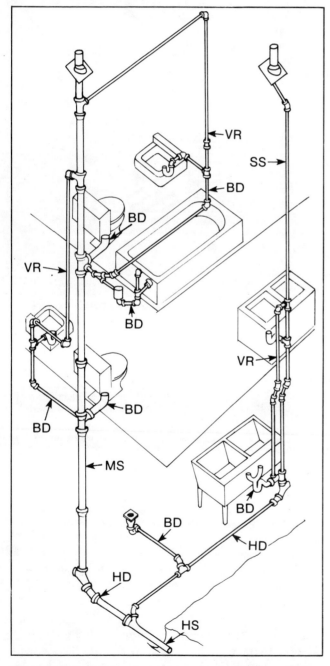

A typical drainage system: MS = main (soil) stack; SS = secondary (soil) stack; BD = branch drain; HD = house drains; HS = house sewer; VR = vent run.

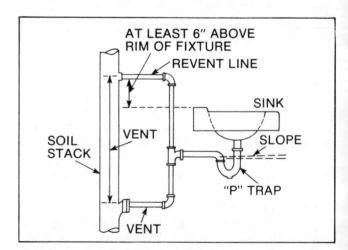

Fixtures are vented to carry off dangerous gases and to permit water and waste to travel freely through waste pipes without backing up into other fixtures.

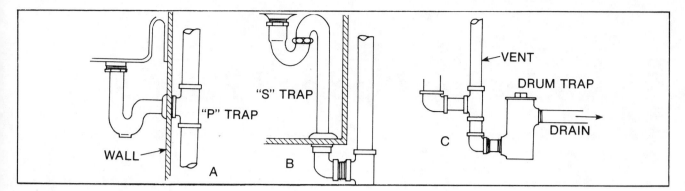

Different types of traps are used with fixtures. The drum trap is normally installed near a bathtub drain with its cover flush with the floor.

Cleanouts Blockage in the waste or soil lines can present a major problem unless you have cleanouts. These are openings in sections of pipe runs, usually at the highest end of the run, and are used for clearing clogs. They are normally made from 45 degree, Y-type fittings with a plug screwed into one branch. It is recommended that every horizontal drainage run have a cleanout. There should be at least one main cleanout at the base of the main stack. The house drain, leading to the sewer, should also have a cleanout. When the line clogs, you unscrew the cleanout plug and insert a drain and trap auger into the blocked section.

Fixtures

The fixtures provide the required means for using your water. In this sense, a faucet on the outside of your house, called a hose bib, is a fixture. So is a laundry tub in the basement, or a shower, dishwasher, or toilet. Each has a purpose connected with your use of water; and each must have certain features to serve its purpose (for instance, the hose bib must be threaded so you can attach the hose; the shower must be designed to mix hot and cold water to give you water of the proper temperature, and so on). A great deal of thought should be given to fixtures as they are generally the most costly plumbing items and should be exactly what you will need.

When selecting fixtures, the first step is to select those that are best suited to your budget and needs. The locations of fixtures in a room, such as in a bathroom, kitchen, or laundry, should be carefully planned for maximum accessibility and convenience.

In addition to the usual bathroom and kitchen fixtures, these are some other popular accommodations to think about:
- Extra bathroom
- Half bath
- Basement toilet
- Basement or garage shower
- Sink for basement shop
- Laundry tub
- Electric clothes washer
- Electric dishwasher
- Electric garbage disposal
- Outdoor hose faucets
- Gardening sink
- Drain in garage

It is important to remember that the costliest single item in a plumbing remodeling job can be the main soil stack. If you cannot use the present main stack, you will have to have a secondary or optional vent. In planning fixture locations, keep the following in mind:
- Every building must have a main soil stack built of 3- or 4-inch pipe (depending on the building code and kind of pipe) from the building drain up through the roof.
- If there is one toilet, it must drain into the main stack. Extra toilets can drain into this same stack if not too far from it. The branch drain for a toilet must be the same diameter pipe as the main stack and must slope properly. If an additional toilet is too far from the main stack, you can provide a secondary (2-inch) stack that is increased to the proper (3- or 4-inch) size throughout the portion through which the toilet will drain. Or, this additional toilet can drain into the main stack if you provide a 2-inch vent at the toilet.
- Other fixtures (a tub, lavatory, sink) can drain into the main stack or a secondary stack. The branch drains must be 1½- to 2-inch pipe, depending on the fixture, and must slope downward.
- Any secondary stack should be a minimum of 1½ inches in diameter and must either be vented out through the roof or connect into the main stack at a point above any branch drain.
- Any branch drain that enters a stack 8 feet or more below another branch drain should be re-vented to prevent siphoning off of the water in the lower fixture trap when the higher fixture is flushed. A re-vent should be the same size pipe as the branch drain.
- All stacks must connect with the building drain, which must be of the same size pipe as the main stack, and must slope down all the way to the final disposal.

Selecting Pipe and Fittings

In the days of lead supply pipes and cast-iron drain pipes, most people had to call the plumber when something major went wrong with the plumbing system. Few would ever consider adding onto the system themselves. Today, however, with the advent of easy-to-work-with pipe materials and fittings, doing your own plumbing is possible for anyone who is willing to learn about techniques, tools, and how to plan a plumbing installation.

Pipes

Steel Pipe Steel pipe is used primarily for carrying water, steam, or gas. Available in both black and galvanized finishes, it is made of wrought iron steel and sold in sizes from 1/8 to 6 inches. Galvanized pipe is commonly used for water systems, black pipe for manufactured and natural gas. Most residential water-supply runs use 1/2-, 3/4-, or 1-inch pipe, while pipes up to 6 inches in diameter are employed for drainage systems. Standard pipe lengths are 10 and 21 feet; however, many plumbing supply dealers will cut lengths to order and thread the ends for you, too.

Galvanized steel is the strongest and least expensive of all supply lines. But, it is generally more difficult to install because of its rigidity and the threading it requires. Use steel pipe if the lines are in danger from tools or heavy objects.

Copper Pipes and Tubing In plumbing, copper is most commonly used in the form of tubing. It comes in two types: hard (or rigid) and soft (or flexible) tubing. There are also five commercial grades of copper tubing: Type K, both hard and soft, is the heaviest and is used generally in commercial work; Type L, both hard and soft, is lighter than K and is popular in residential water lines; and Type M is made in hard tubing only and is used for light residential lines. Your local plumbing code should be checked before installing Type M.

Most rigid tubing in the home is in 3/8-, 1/2-, and 3/4-inch sizes for in-house use (Type L) and in 3/4-inch and larger sizes (Type K) for underground use; all are sold in lengths up to 20 feet. Thin-walled, rigid tubing (Type M) comes in 3/8- to 1-inch sizes for interior use only. Solder-type fittings are always used and are very easy to assemble once you have learned how. By using the proper adapters, you may combine copper and threaded steel pipes in your installation.

If you heat rigid copper tubing, you remove its temper. To form curves, use a propane torch to warm a section of the pipe, then bend it by hand, using the techniques described below for soft copper tubing.

Since copper tubing has a smooth bore, water flows through it with less resistance than through wrought iron. This feature permits replacing a heavy iron pipe with a copper tube of smaller diameter. To determine possible replacement sizes involving this factor, check the table. For distances over 20 feet, or for risers running up through the walls, it is safer to use a copper tube of the same diameter as the iron pipe it replaces.

SIZES OF TYPES K, L, AND M COPPER TUBING				
Nominal Size in Inches	Outside Diameter	Inside Diameter		
	Types K-L-M	Type K	Type L	Type M
3/8	.500	.402	.430	.450
1/2	.625	.527	.545	.569
3/4	.875	.745	.785	.811
1	1.125	.995	1.025	1.055
1 1/4	1.375	1.245	1.265	1.291
1 1/2	1.625	1.481	1.505	1.527

Soft or flexible copper tubing is the easiest metal pipe to install. Its walls are thinner than rigid copper and so the cost is less. It comes in 60-foot coils and is available in the same two types (L and K) and the same sizes as rigid copper tubing. Soft copper tubing has the advantage of turning corners without fittings. This means you can avoid complicated joints to round corners. Just bend the pipe around them. The only trick is to keep the tubing from sinking, flattening or pinching while you are bending it. If the diameter of the tubing is reduced, the flow of water through the pipe will also be reduced. Unless a special tube-bending tool is available, you should make the curves as gradual as possible. One method is to lay the tubing on a board, fastening down one end, then kneeling on the tube and raising the free end slowly. Move the knee toward the free end slightly, and raise again, repeating until the desired curve is obtained.

Another way of bending soft copper tubing requires a bending spring, an inexpensive device that is slipped down the pipe to the proper location. Grip the pipe firmly, placing your knee in the middle of the spring, and pull both ends until the desired bend is produced. The spring arrangement prevents kinking. Soft copper tubing can be bent by hand.

Soft copper tubing can be assembled with either solder-type fittings or flare-type fittings. The latter

PIPE DATA AT A GLANCE

Type of Pipe	Ease of Working	Water Flow Efficiency Factor	Type of Fittings Needed	Manner Usually Stocked	Life Expectancy	Principal Uses	Remarks
Brass	Threading required or ask for pre-threaded. Cuts easily, but can't be bent. Measuring a job rather difficult.	Highly efficient because of low friction.	Screw-on connections.	12-ft rigid lengths. Cut to size wanted.	Lasts life of building.	Generally for commercial construction.	Required in some cities where water is extremely corrosive. Often smaller diameter will suffice because of low friction coefficient.
Copper Pipe	Easier to work with than brass.	Same as brass.	Screw-on or solder connections.	12-ft. rigid lengths. Cut to size wanted.	Same as brass.	Same as brass.	
Copper Tubing, Rigid	Easier to work with than brass or hard copper because it bends readily by using a bending tool or by annealing. Measuring a job not too difficult.	Same as brass.	Solder connections.	3 wall thicknesses: K-thickest L-medium M-thinnest. 10-or 20-foot lengths.	Same as brass.	"K" is used in municipal and commercial construction. "L" is used for residential water lines. "M" is for light domestic use only: check codes before using.	
Copper Tubing, Flexible.	Easier than soft copper because it can be bent without a tool. Measuring jobs are easy.	Highest of all metals since there are no nipples, unions, or elbows.	Solder or compression connections.	2 wall thicknesses K-thickest L-medium 30-, 60-, or 100-foot coils (except "M").	Same as brass.	Widely used in residential installation.	Probably the most popular pipe today. Often a smaller diameter will suffice because of low friction coefficient.
Galvanized Steel (or Wrought Iron)	Has to be threaded. More difficult to cut. Measurements for jobs must be exact.	Lower than copper because nipple unions reduce water flow.	Screw-on connections.	10- or 21-foot rigid lengths. Usually cut to size wanted.	Very durable.	Generally found in older homes.	Recommended if lines are in a location subject to impact.
Plastic Pipe	Can be cut with saw or knife.	Same as copper tubing.	Insert couplings, clamps; also by cement. Threaded & compression fittings can be used. (Thread same as for metal pipe.)	Rigid, semi-rigid & flexible. Continuous lengths to 1000 ft.	Long life & it is rust & corrosion-proof.	For cold water installations. Used for well casings, septic tank lines, sprinkler systems. Check codes before installing.	Lightest of all, weighs about 1/8 of metal pipe. Does not burst in below freezing weather.

type hold simply by being tightened and are very easy to assemble. However, flare-type fittings should never be used inside walls or anywhere they cannot easily be reached.

Soft tubing is easy to work with but is not as durable or professional-looking on exposed runs as rigid tubing. On the other hand, rigid tubing is more difficult to cut and bend. You can employ soft copper tubing on hidden plumbing runs and rigid tubing on exposed runs, such as in the basement.

In older houses, you might come across hard copper pipes, not to be confused with rigid copper tubing. Copper pipe uses the same sizing system and the same style fittings as galvanized steel or iron pipe; however, do not connect copper to iron or steel unless you use a fitting specially designed for that purpose or wrap the threads with plastic joint tape. Without the special fitting or tape, certain conditions cause an electro-chemical reaction to occur between the two metals, corroding the joint and eventually causing a leak.

Rigid and Flexible Plastic Pipe Rigid plastic pipe in 10-foot lengths and flexible plastic pipe in 60-foot coils are similar in sizes and uses to copper tubing. The CPVC (chlorinated polyvinyl chloride) pipes can be used in most water applications except that they are not made for hot-water systems with pressures over 100 pounds per square inch at temperatures over 180 degrees F. And, being nonconducting, they cannot be used in grounding. In recent years, PB (polybutylene) and CPVC have gained code approval for hot-water systems.

Plastic pipe hardware weighs only one-eighth as much as iron pipe and one-third as much as copper tubing. The plastic surface prevents interior scale buildup and corrosion. Also, the natural insulation property of plastic reduces condensation on the outside of cold-water lines. In addition, because of this natural insulation, the pipe can be installed just a few inches under the surface of the soil since freezing weather has no adverse effect on it. But, if the water is to be used for drinking, the pipe should bear the seal of the National Sanitation Foundation.

The plastic pipe can be connected to existing iron or copper piping with plastic-to-metal connectors and couplings. When joining plastic to an iron pipe system, try to use the lines where a connection is fairly close. This way you can disassemble the existing pipe from the connection and back to an existing elbow. A new iron tee replaces the elbow for the new connection. In copper lines, connections can be made almost anywhere. A CPVC take-off tee is used with copper tubing adapters. But before you start, check your local plumbing codes since some areas have not yet updated codes to permit use of plastic pipe. (Plumbing ordinances, as a rule, permit either galvanized steel or copper pipe for water-supply purposes.)

The industry's technology is moving fast these days, so be sure to consult your dealer about any new ad-

vances that may prove beneficial to you and your plumbing project.

Fittings

Many attachments, connections, turns, etc., are necessary to install pipe in the proper location and terminate it at the right spot. Although each kind of pipe requires its own fittings, there are certain similarities in all fittings. There are, however, two basic designations for pipe fittings: male and female. These refer to the threading. Male threading is on the outside and threads into the female threading which is on the inside of the fitting.

Steel Pipe Fittings

Pipe fittings are stocked in many sizes and shapes and for many purposes. Order them according to the pipe size on which they will be used. (A ½-inch elbow will accept ½-inch pipe at both ends.) Here is a brief rundown on the popular steel pipe fittings.

Nipples Used to extend a line or provide proper threading at the right location. Nipples come in diameters ranging from ⅛ inch to 4 inches to match all standard pipe diameters and in lengths from close (nipples that are threaded on both ends to a point where threads almost join in the center) through 24 inches. Normal size increments are even inches. Long nipples of "cut lengths of pipe" which are threaded on both ends are available in about 24-inch lengths, usually increasing in length by 6-inch increments (30 inches, 36 inches, 42 inches, etc.).

Couplings Connect all standard sizes of pipe. Tight seal with a pipe wrench and pipe joint compound will waterproof connection.

Bushings Inserted inside a coupling to reduce the size of the pipe. With a coupling, a run of pipe can be reduced a size or two; with a series of bushings, any number of reductions in size of pipe can be made.

Floor Flanges Connect pipe to a wall, floor, or any flat surface. Flanges are threaded onto pipe and tightened. This provides a flange rim with four screw holes, making it easy to affix pipe to any flat surface.

Elbows Change direction of pipe. Most common are 90-degree and 45-degree elbows which have inside threads on both ends. A street elbow has inside thread on one end and outside thread on the other.

Reducers Like bushings, serve to reduce pipe size. Bushings screw into a coupling while reducers screw directly onto pipe threads. Some reduce pipe only one size; others can reduce several sizes.

Side Outlet Elbows Have three-way outlets. Can be used as corner pieces for railings, fences, etc. Also used for pipe connections on corner construction.

Crosses and Tees Tees are available in all sizes and shapes. Most common is the straight tee which has three inside threads of the same size which can be used

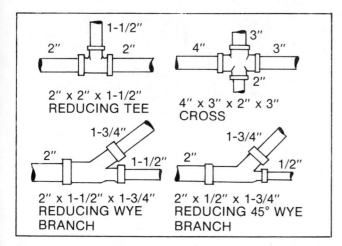

How to order reducing fittings.

to run three pipes in a "T" shape. Reducing tee has same shape, but two straight ends of the "T" are reduced one size or more.

Four-Way Tee (Side Outlet) Similar to the side outlet elbow except side outlet runs through the elbow with an opening of the same size on each end. Straight cross has four outlets for pipe of the same size. Side outlet cross has an opening on the side for a fifth pipe of the same size.

Return Bends and Y Bends Return bends are made in close, medium, and open patterns. The close is a sharper bend than the medium and the medium is a sharper bend than the open. The Y bend is a straight or reduced outlet which permits affixing pipe of the same or reduced size to a 45-degree angle.

Ground Joint Union A three-part item used to connect any standard size pipe where it may be necessary to disconnect later. Because of the bronze-to-bronze or bronze-to-iron ground joint seat, this item can readily be taken apart and reassembled at the nut with only a pipe wrench and no pipe joint compound.

Copper Pipe Fittings

Fittings for copper pipe must be soldered on at least one end, leaving one or both ends unthreaded. After flux has been applied, solder is introduced at the edge of the fitting. It is then drawn, by capillary action, the full depth of the fitting and completely surrounds the tube. The result is a strong, leakproof, bonded joint. The solder is usually applied with a propane torch, a process known as "sweating" fittings.

Flare and compression fittings are also used to make copper tubing connections. Appliances that use a small amount of water, such as ice maker refrigerators, evaporative coolers, and humidifiers, use a small size slip joint tee and saddle tee. Slip joint tee is installed by cutting the line and spreading it slightly. The saddle tee clamps onto the line and a hole is drilled in the pipe through the side opening.

A commonly used fitting in installing dishwashers

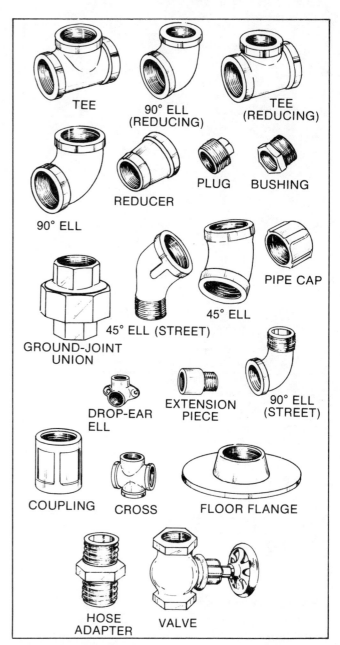

Many types of fittings are available for use with galvanized steel pipe.

is the three-way compression stop. This fitting, installed on the hot or cold sink supply line, will cut the flow of water on both appliances at the same time.

There are special fittings available for making copper to steel connections.

CONVERTING STEEL PIPE TO COPPER TUBING	
Iron pipe size (inches)	Copper tube size (inches)
½	⅜
¾	½
1	¾
1¼	1
1½	1¼
2	1½

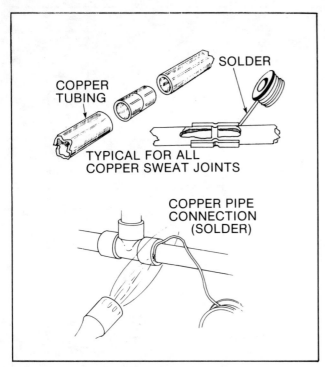

Copper pipe connection with solder.

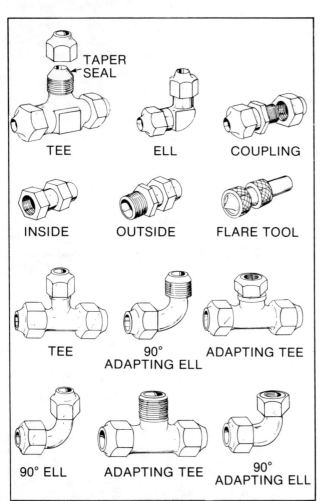

When working with flexible copper tubing, fittings may be soldered or you can use the flare-type.

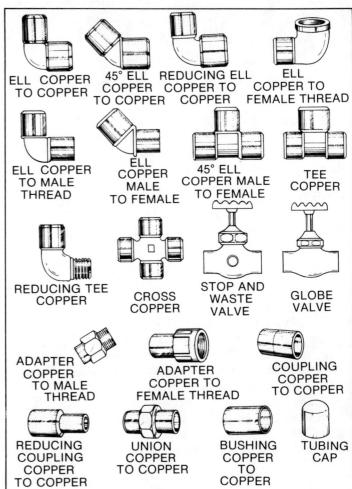

Soldered fittings are used on rigid copper tubing.

Fittings for Plastic Pipe

Plastic fittings for plastic pipe, metal fittings for plastic pipe, and fittings for joining metal and plastic pipe are available in threaded, insert, and compression types and are generally the same as those for metal pipe. In fact, threaded fittings thread exactly like metal fittings. Insert fittings are inserted into the pipe and held in place by a steel clamp. Other plastic and metal fittings can convert thread fittings to insert fittings or insert to thread.

Plastic fittings can also be connected with special cement or solvent which is brushed on with a regular paint brush, dries in about 5 minutes, and forms a permanent joint. Plastic pipe can also be connected to steel pipe by using special fittings.

Valves and Faucets

Valves are used in water-supply lines to shut off the flow of water when desired. A faucet is a valve used at the end of a water-supply line.

Metal valves in home plumbing lines usually are cast bronze and have portions machined and threaded for trimming. The more common valves found in a residential water-supply system are as follows:

Gate Valve It has a sliding wedge that is moved across the waterway, usually by a threaded spindle or stem. It is either rising or non-rising, the latter having a shorter bonnet.

A gate valve is used to completely shut off or open a waterway, but not to control the volume of flow. Either opening of a gate valve may face the pressure side of the line. Gate valves allow complete passage of water without causing additional resistance to the flow and should be used on supply lines that are in constant use.

Globe or Angle Valves They are used when a valve must be opened and closed frequently and water pressure is high. Globe valves are often used to control volume of flow. They have two chambers with a partition between them for passage of water which must change course several times from port to port, introducing resistance. Therefore, globe valves should not be used in water-supply lines to serve occasional shut-off purposes only.

An angle valve is similar to a globe valve, but has its ports at right angles. Water passage is larger than on a globe valve and since there is only one change of direction of flow, less resistance is introduced. An angle valve installed at a turn in piping eliminates the necessity of an elbow and is often preferred to using a globe valve and elbow.

Plug and Key Valves They are better known in the plumbing industry as straight stops. They have tapered ground plugs that seat into matched tapered ground bodies. On top, the plugs have flatheads, squareheads, and socket-heads; the lower end is threaded to hold a hex nut and friction ring combination. This is mounted over a tension spring inside the body that keeps the plug tight. They are manufactured in brass, bronze, galvanized iron bodied, and black iron bodied and are used mostly as gas stops.

Drainable Valves or Stop and Waste Valves They have a small opening on their non-pressure side to allow drainage when it is in the cut-off position. Sometimes it is called a bleeder valve and may be obtained in threaded, sweat, flare, and slip joint ends, the latter two municipal and emergency, respectively. They are manufactured mostly in the flathead and socket-head styles. The residential types with a socket-head usually take a ⅜-inch key rod.

Check Valves They operate automatically, permitting flow in one direction only. They are sometimes combined with a throttling or shutoff valve. Some communities require a check valve in a cold-water line between the water heater and meter.

Check valves are used to prevent water pumped to an overhead tank from flowing back when the pump stops. Some check valves are designed for use on vertical pipes only. Therefore, correct installation is essential. The closing device—disk, ball, or clapper—would fall shut by gravity when installed vertically.

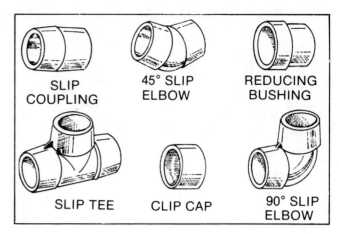

These fittings for plastic pipe are "welded" on with a special cement or solvent.

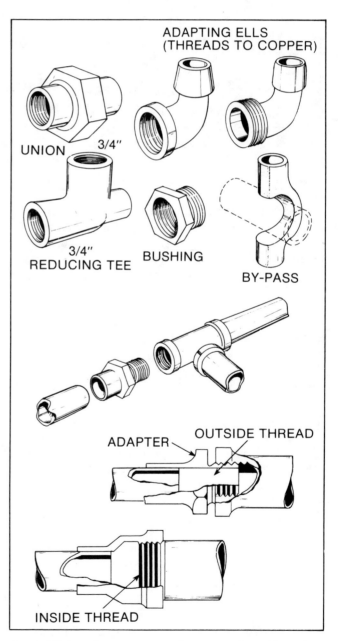

Special fittings for copper to steel connections and how the connection is made.

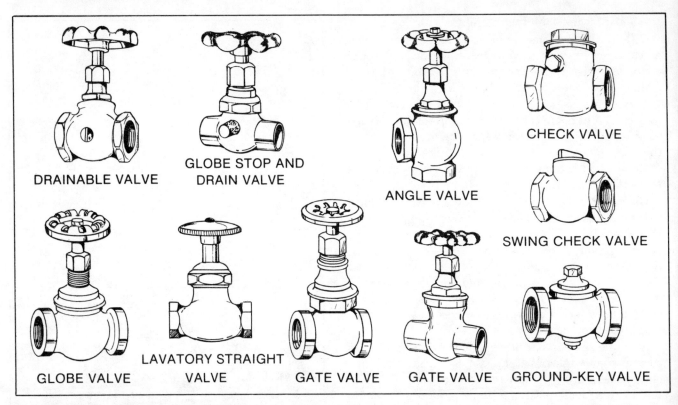

DRAINABLE VALVE

GLOBE STOP AND DRAIN VALVE

ANGLE VALVE

CHECK VALVE

SWING CHECK VALVE

GLOBE VALVE

LAVATORY STRAIGHT VALVE

GATE VALVE

GATE VALVE

GROUND-KEY VALVE

Typical valves commonly found in the home's water-supply system.

Swing-Type Check Valves They are used to prevent water pumped into an overhead tank from flowing back when the pump stops. A small, smooth swing-type gate is located in the center of the valve. As water is pumped through the flow side of the valve, the gate swings open to allow the passage of water. If water attempts to back-up through the valve, the gate is forced shut against the pressure side of the valve.

Plastic Valves There are two basic types of plastic valves. One type is threaded for use in hot- and cold-fluid systems, and the other is non-threaded solvent-weld for use with existing plastic piping systems. Both types are available with or without a washer.

For use with hot- and cold-fluid systems, the plastic valves include globe (stop), stop and waste, boiler drain, flanged angle sill cock, and sink faucet valves. The valves perform with a temperature of -20 degrees F to 180 degrees F. They are made for pressures up to 150 pounds and are excellent for hard water areas, as they resist mineral build-up.

Since handles are made of plastic, heat is never absorbed from the water line. Do-it-yourselfers will enjoy the ease with which they can be installed. No brute force is required. Because the molded threads are more "perfect" than machined threads on a metal part, the installer can mate the parts one to two threads beyond normal make-up on a metal joint for a better connection.

A "double seal" feature allows the washer to be removed and the plastic seat alone will maintain the integrity of the valve. In some environments, where it is felt the washer or metal parts may deteriorate, it is recommended that installation be made without the washer.

Solvent-weld (CPVDC) valves for hot- and cold-fluid systems or PVC pressure piping systems offer the same benefits as the above type. They are available only in globe (stop) type.

Faucets There are two basic types of faucets on the market today: the washer type and the washerless type. The washer type, also called a stem or compression faucet, is the one found in most older plumbing systems and is the least expensive faucet to buy. The washerless or noncompression faucets are considered the modern day type which use diaphragms, balls, cartridges, disks, and valves in place of washers. These "new" faucets are more costly, but less troublesome than the washer type units. They are slowly replacing the washer faucets as standard in modern plumbing systems.

Fixture Supply Lines Most exposed fixture supply lines are chrome-plated for best appearance. Available in rigid and flexible types, they are equipped with proper fittings to join fixtures to their supply piping and are best ordered with the fixtures to guarantee correct fit. Supply stops are often used as shutoff valves between water pipes and fixture supply lines.

Traps The function of a trap is to prevent air from entering the waste pipe while liquid is flowing, and thus it eliminates noise. It also stops unpleasant odors from entering the house from the sewer lines. Remember that many common sewer gases not only are ob-

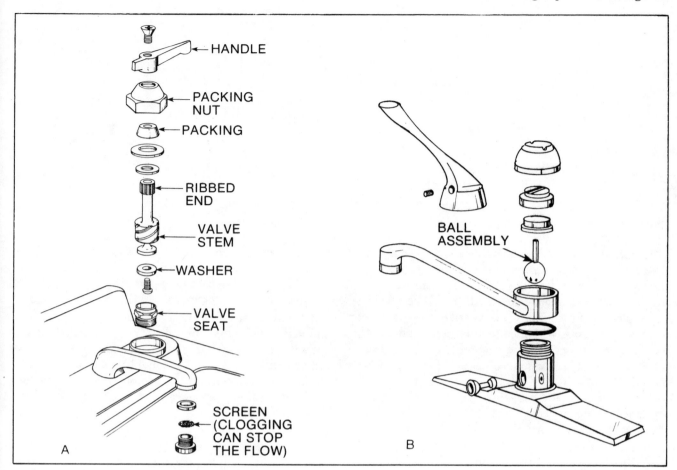

(A) Parts of a typical washer-type faucet. (B) Parts of a typical washerless-type faucet employing a ball assembly.

noxious, but can cause serious illness and even death.

The two most popular types of traps are "P" and "S" traps. When working in tight places, you can install "P" and "S" traps somewhat off the center line to avoid cutting into studs or joists. All fixture traps in the house must be constructed with a cleanout opening at the bottom of the curve so that waste collections can be removed when necessary.

Like exposed fixture supply lines, all exposed fixture drains and traps should be of chrome-plated brass.

Drainage System Pipes

There are three types of pipe used today for drain-waste-vent (DWV) plumbing systems: (1) cast-iron and threaded steel pipes; (2) plastic pipes; and (3) copper pipes.

Cast-Iron and Threaded Steel Pipes Cast-iron pipe has long been used for stacks, building drains and sewers, and any drains buried underground. It comes in 2- and 4-inch sizes in 5-foot lengths, either with a hub at one end or with hubs at both ends (for making two shorter hub end pieces), and in two weights: heavy weight and service weight. Service weight is lighter and easier to handle; use it if your local building codes allow.

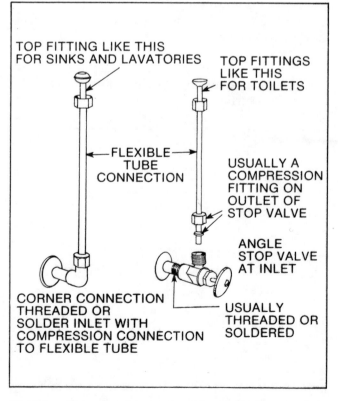

Flexible tubing carries water from the supply pipes in the wall to individual fixtures.

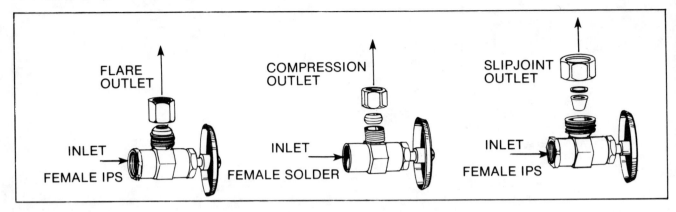

Supply stops are useful as shutoff valves between water pipes and fixture supply lines.

In recent years, no-hub or hubless cast-iron pipe has become popular for drainage installation, especially for the do-it-yourselfer. The no-hub sections join with a neoprene rubber gasket held with a ribbed stainless steel sleeve and clamps. It is particularly useful for connecting new DWV lines to older lead-and-oakum-caulked cast-iron drainage systems.

No-hub cast-iron pipe comes in 5- and 10-foot lengths and 2-, 3-, and 4-inch inside diameters. A 3-inch no-hub pipe fits inside a 2-by-4 inch wall with no special boxes or furring. It may be used vertically or horizontally, above or below grade.

Threaded, galvanized steel pipe, available in 1½- and 2-inch sizes in 21-foot lengths, is generally used with cast iron for branch drains, vent lines, and sometimes for secondary stacks. Steel pipe, however, must not be buried underground.

So-called sanitary fittings must be used for assembling all drainage lines. Sanitary fittings have recessed threads that make a smooth connection through which water flows unobstructed. They are installed facing one direction only. With an ordinary fitting, the pipe ends protrude into the line and could snag solid wastes, causing clogging. Water-supply lines use ordinary fittings; vents can take either ordinary or sanitary. When used for vents, sanitary fittings should be installed upside down to lead the gases up and out.

Plastic Pipe Plastic DWV piping is made in two materials: polyvinyl chloride (PVC) and acrylonitrile-butadiene-styrene (ABS). While both materials are safe for all the usual household wastes, some local plumbing codes require the use of ABS where plastic drainage systems are permitted. Both ABS and PVC pipe and fittings come in 1½-, 2-, 3-, and 4-inch diameters and 10-foot lengths.

Being light in weight, plastic DWV pipe is easy to handle. The fittings are all uniform and exact in dimension, as is the pipe itself. Each material has its own solvent cement to make all the joints. The two materials may be joined to each other only if you use ABS solvent—the PVC cement will soften ABS pipe. Incidentally, all plastic DWV fittings are of the sanitary type and are used inverted wherever needed for vent-run connections.

Copper Pipe Lighter and easier to install than cast iron, copper pipe and fittings are available in 1½-,

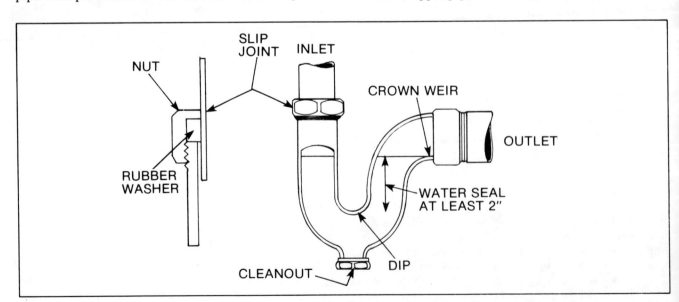

Typical "P" type trap assembly.

2-, and 3-inch sizes (pipe in 10-foot or 20-foot lengths) for all in-house drainage and vent runs. You can also buy it cut to order. It comes in two weights: Type M and DWV. DWV is lighter in weight, easier to use, and recommended if codes permit. Drainage-type copper must not, however, be used underground. All fittings are the sanitary type and are used inverted where needed in vent runs.

Outdoor Drainage Lines Fiber, vitrified clay, concrete, and asbestos-cement are materials used for the portion of your drainage system located outside the house. Your local code will specify the type to use if you are connecting to a municipal system or to a septic tank or drain field. In addition to plastic and cast-iron pipes, there are other popular drainage pipes you can use.

Fiber pipe is more expensive than some others, but is a time-saver to install. It is usually stocked in 8- and 10-foot lengths, making it much easier to maintain a downward pitch than when using vitrified clay or cement tile types.

Vitrified clay is noted for a smooth interior surface, ease of installation, and permanency. It is stocked usually in 2- to 3-foot lengths, either straight or curved. Cut it the same as cast-iron pipe and close the joints with cement or special compounds.

Concrete pipe is generally stocked in 2- to 4-foot lengths and is joined the same as vitrified clay pipe. Acids have more effect on concrete pipe than on vitrified clay.

Asbestos-cement pipe is both extremely durable and resistant to corrosion. It is usually stocked in 5- and 10-foot lengths and is highly recommended for sewer lines.

Converting Your Plumbing Ideas to an Installation Plan

To help order the necessary pipe and fittings, you must make an "installation plan." The illustrated plan is for a typical cast-iron drainage system. The part labeled "Elevation" is a side view to show the vertical "runs" of pipe, and the part labeled "Plan" is a top view

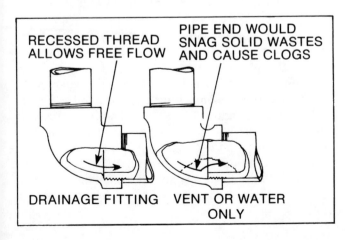

RECESSED THREAD ALLOWS FREE FLOW

PIPE END WOULD SNAG SOLID WASTES AND CAUSE CLOGS

DRAINAGE FITTING   VENT OR WATER ONLY

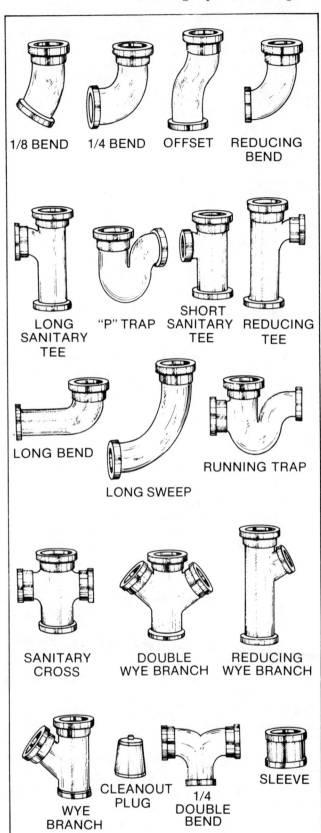

1/8 BEND   1/4 BEND   OFFSET   REDUCING BEND

LONG SANITARY TEE   "P" TRAP   SHORT SANITARY TEE   REDUCING TEE

LONG BEND   LONG SWEEP   RUNNING TRAP

SANITARY CROSS   DOUBLE WYE BRANCH   REDUCING WYE BRANCH

WYE BRANCH   CLEANOUT PLUG   1/4 DOUBLE BEND   SLEEVE

Typical drainage fittings.

Comparison of sanitary drain fitting (left) and vent or water fitting (right).

(looking down) to show the horizontal runs. Wherever necessary, details should be shown in an enlarged boxed-in view arrowed to the point where the detail fits in (see number 1). The large numbers (such as 1 through 7) in the illustration give you the installation steps. All pipes and fittings should be labeled for ready identification. You should make similar plans for a water-supply system, too.

When making your plans, the following are some factors that should be taken into consideration:

1. If a new main soil stack must be added, its location should be chosen first. This governs all other drainage piping requirements, and the water-supply piping can easily be adapted to any location made necessary by the soil stack. This stack must be straight, if possible, or have only gentle bends (if the building layout makes it necessary to have bends). It must be located directly behind the toilet or as close to it as possible. If there are two toilets, it should serve both, or a second stack is required. In this case, first consideration is given to the main stack, which will serve one toilet and most of the other fixtures. The second stack may be like the main stack or may use smaller (2-inch) pipe above the toilet drain. Any stack must have drainage fittings in the portion that carries drainage; but use ordinary fittings in the vent portion (that carries only air and gases). In very cold climates, outer walls should be avoided as pipes located in them may freeze. The bottom of any soil stack should be in a position where: (A) a cleanout can be conveniently located; and (B) the stack can easily be connected to the building drain.

2. Once the stack is located, the drainage lines should be placed. These lines must be sloped correctly, and each must emerge from the wall or floor where proper connections to the fixture can be made.

3. Re-vents should be located next (if used). These have the same requirements as drainage lines, except that ordinary fittings are used and they do not slope.

4. Any secondary soil stacks are now located. These have the same requirements as the main soil stack.

5. The building drain is the last part of the drainage system to be planned as its principal requirement is to collect from the soil stack(s) and drain to the final disposal. The sewer line must slope downward (all the way to final disposal) at the proper pitch.

6. Once the drainage system is planned, the water-supply system can be planned to keep all the pipes running (as nearly as possible) through the same wall openings (to save installation work). Your furnace (and chimney) location will prob-

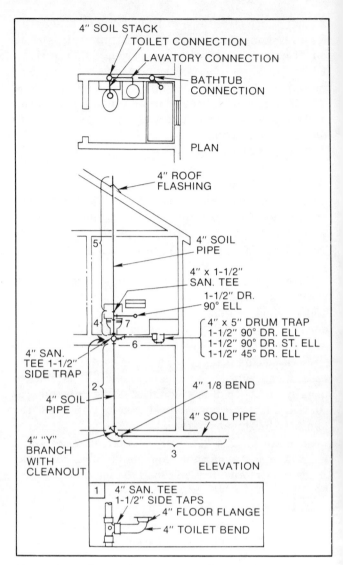

Converting a house plan to an installation plan.

ably determine the hot-water heater location (unless your heater is an electric one), and hot water pipes to various fixtures should parallel cold-water pipes to these fixtures (for convenience and the appearance of the installation). The point at which your water-source line enters the building will determine the meter (if any) location (as it must be at this point); and all pipes in the building should be planned to slope downward to a low point (as near as possible to the meter) so that you can locate a stop-and-waste valve at this low point.

Pipe Sizes A faucet is turned on somewhere in the house and the person taking a shower finds that the mixture of hot and cold water suddenly becomes too hot or too cold. One possible reason for this common complaint is that the original installation of pipe was faulty. Plan to use a large enough pipe so that water pressure to the fixture is not affected when other fixtures are turned on. Pipe sizes are recommended as follows:

NOMINAL SIZES FOR FIXTURE SUPPLY, DRAIN, AND VENT LINES

Type of fixture	Lavatory	Tub or shower	Toilet	Sink	Garbage disposal	Dish-washer	Clothes washer
Fixture supply lines Branch lines	⅜-inch	½-inch	⅜-inch	½-inch		½-inch	½-inch
Fixture drains Branch drains Re-vent Lines	1½-inch	1½-inch	3- to 4-inch	1½- to 2-inch	1½- to 2-inch	1½- to 2-inch	1½-inch

NOTE: Main soil stack: 3-4 inches. **Secondary soil stack:** Size of largest branch drain connected to it, in most cases. **Basement floor drain:** 2-4 inches. **Building drain:** At least the size of main soil stack. **Branch building drain:** At least the size of largest secondary soil stack emptying into it. **Vent:** Same size as drain pipes; some codes permit 2-inch venting if it serves no other fixture. **Cold-water main line** serving both the cold-water system and the hot-water heater: ¾-inch to 1-inch. **Cold- and hot-water main lines** serving two or more fixtures: Size of largest branch line served or, if fixtures will be used simultaneously, the next pipe size larger than that used for the largest branch line.

Drainage pipe capacity is usually based on the fixture unit figures given in the table below. A fixture unit is the National Plumbing Code term used as the basis for figuring DWV pipe sizes. It represents a waste flow of one cubic foot a minute (about 7½ gallons). After checking on drain pipe diameter, size the main stack. If only one or two toilets will empty into its base, a 3-inch stack is plenty large. Add up the fixture units for all house fixtures to see how large the main building drain must be to handle the load.

Remember that no drain-waste-vent system is truly horizontal. Drainage line must be pitched downward at least ¼ inch per foot in order to drain. On the other hand, a vertical drain pipe is really vertical.

Planning Plumbing Improvements

Plumbing costs can often be kept down by good planning in locating fixtures. Fixtures located back to back on opposite sides of a wall save on piping. Locating all bathroom fixtures on one wall, as shown in the illustration, also saves piping. In the arrangement shown, one vent stack serves all fixtures. A vertical arrangement of fixtures can reduce the amount of piping needed in multi-storied houses. Locating fixtures in a continuous line saves piping in single-story houses.

A water heater should be located as close as practical to the fixture where hot water will be used most frequently. Long runs of hot-water pipe result in unnecessary use of water and heat.

When adding a bathroom, your major plumbing problem will be how best to provide a soil stack for the new fixtures and how to connect this stack to your existing building drain. Of course, if your new room will butt against a wall in which there already is a suitable size soil stack, this problem is solved. More often, however, you will need to add a new soil stack. The easiest place to put this is inside one of your new walls, preferably the one closest to your building drain.

Fixture units

Fixture	Fixture units
Toilet	4
Bathtub/shower	2
Shower only	2
Sink	2
Lavatory	1
Laundry tub	2
Floor drain	1
One 3-piece bathroom group (toilet, bathtub, lavatory)	6

Pipe capacity

Pipe size	Fixture units permissible	
	Horizontal pipe	Vertical pipe
1½"	3	8
2"	6*	16
3"	20**	30***
4"	160	240

*waste only
**not more than two toilets on horizontal line
***not more than six toilets on one stack

If you are creating a new bathroom within your present house walls by building three (or two) new partitions to enclose an area (like the space under a stairway or in the corner of a large room), then you may be able to install a new soil stack inside one of the new partitions. In a one-story house this is easy to do as the stack can open right out through the roof for venting. If there is another floor above, however, you will have to run the stack up through a partition on the floor above, or plan to box it in. If this is impossible, you may be able to run the new stack up an outside wall, on the outside, and box it in to match the house exterior. In this case, the stack should be insulated against freezing.

When modernizing old homes, plan to locate plumb-

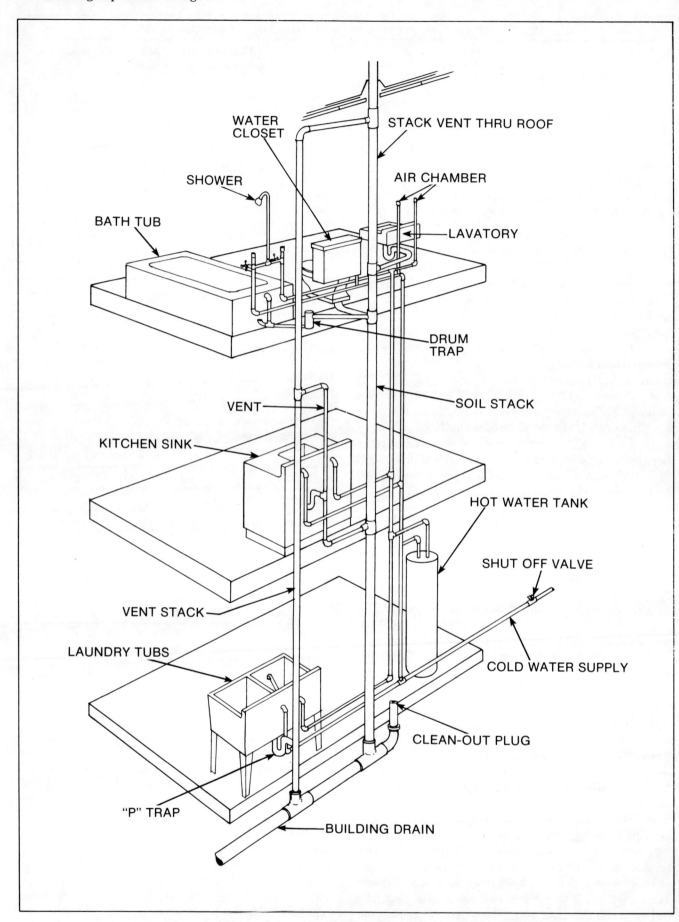

Plumbing arrangement in a two-story house with a basement.

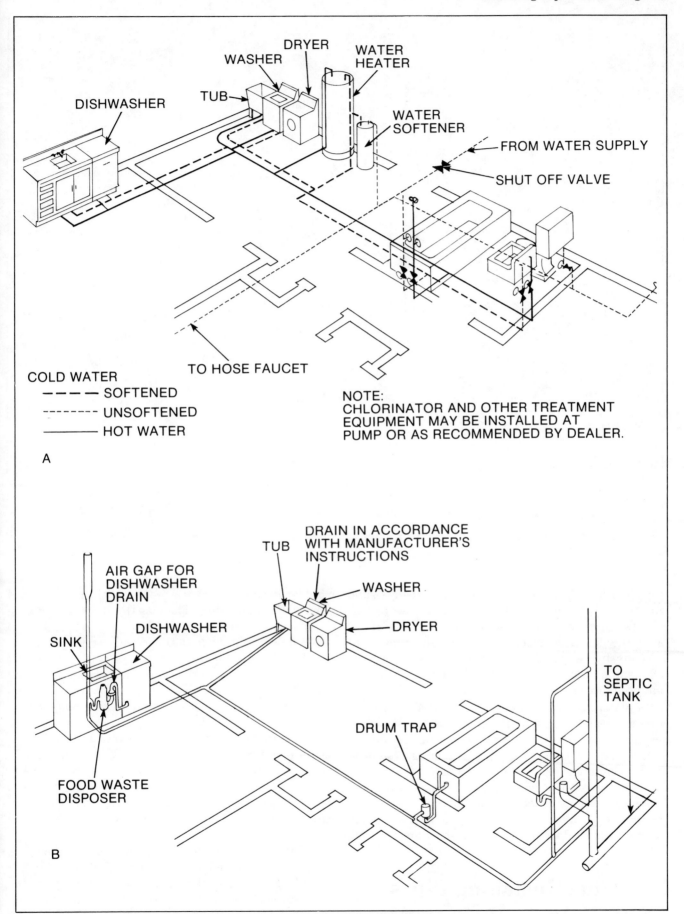

(A) A fixture and water-supply-piping layout for a one-story house. (B) Drainage system layout for the arrangement in A.

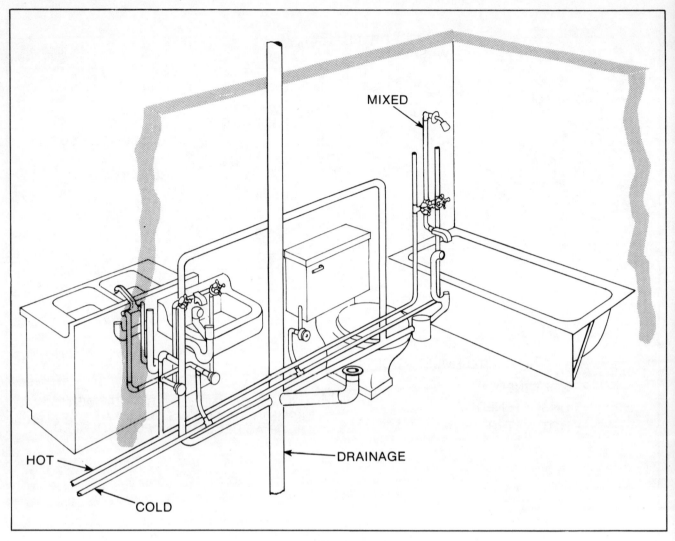

Plumbing fixtures located back to back on opposite sides of a wall.

ing facilities relating to the kitchen, bathroom(s), and the laundry as close together as possible since this makes the job much easier and less costly. Also try to keep the hot-water pipes and drains with their vent pipes as short as possible. This not only makes for more efficient plumbing but makes the installation and maintenance costs less.

In planning a plumbing system, consider your future needs as well as your present needs. It costs less to install a few extra tees with plugs for future connections than it does to cut into a plumbing system to make connections later on. Adding or remodeling plumbing in existing buildings involves the additional expense and labor of opening up walls or floors. It may be more economical to run piping along the exposed face of a wall or floor and then box it in for appearance.

When Installing Pipes

There is, of course, no problem in installing pipes in a house under construction; all pipes that are to be hidden in walls are positioned before the inner sides of the walls are finished. Then, again, if you are making your installation in an old building, but intend to assemble the piping in the open and box it in later, you will not have to break through very much of the walls and floors. If, however, you plan to conceal all piping inside the existing walls and floors of an old house, you will have to remove some wall covering and flooring and will probably have to cut through some joists (horizontal beams that support floors) and studs (vertical partition supports). "Breaking through" chiefly concerns drainage pipes because these are large and cumbersome. Water-supply pipes create no problems, as they are usually run in the same spaces with drainage pipes. Careful planning will eliminate unnecessary work.

Clearance Needed for Pipes There must be sufficient space inside a partition, or a floor (if there is a ceiling below), for the pipes to be run through them. Measure the clear (air) space inside. Space requirements for drainage pipes are shown in the following table.

Pipe size (inches)	Cast iron (inches)		Copper (inches)		Plastic (inches)		Galvanized Steel (inches)	
	Pipe	Fittings	Pipe	Fittings	Pipe	Fittings	Pipe	Fittings
1½			2	2⅛	3	3¼	3	3½
2	4	4	3	3⅛	3½	3¾	3½	4
3	5½	5½	3½	3⅝	4½*	4¾		
4	6¼	6¼	4½	4⅝	5½	5¾		

SPACE REQUIREMENTS FOR DRAINAGE PIPES

* Thin-wall pipe requires 3½ inches.

If you know the size of your studs (or joists), you can figure space as follows: A 2 by 4 stud partition has approximately a ¾-inch clearance inside it. It will not take even a 2-inch cast-iron pipe with hubs, but will take a 2-inch no-hub iron pipe or a 3-inch copper or plastic pipe with fittings. A 2 by 6 stud (or joist) has approximately a 5¾-inch clearance. It will take a 2-inch hubbed cast-iron pipe. A 2 by 8 stud has approximately a 7¾-inch clearance. It will take up to a 4-inch cast-iron hubbed pipe. Because plastic and copper pipes are easily cut to any length desired, fittings can be located as desired. Therefore, you can run 1½-inch pipe (without fittings) through a space as small as 1¾ inches, or 3-inch pipe (without fittings) in a space of only 3¼ inches. Fittings are then planned to be where more space is available. Also, no turning space (for tightening) is required with plastic or copper, as it is with any threaded pipe.

Planning Space for Pipes If a partition is too "thin," the easiest method of thickening it is to set 2 by 4 studs against the existing studs after the walls have been opened and the plumbing installed. Then refinish your wall surface. If an outer wall is being used, and you cannot break into it because the plaster is against the brick or other material used for the outer wall construction, you may have to add 2 by 6, or even 2 by 8 studs for sufficient thickening.

Since each stud in a partition shares its portion of the load, careless weakening of any stud is poor practice. Note how the studs are notched and reinforced, and follow these rules:

- Do not notch the lower half deeper than one-third without reinforcing it with a steel strap or furring.
- Do not notch the lower half deeper than two-thirds, even if it will be reinforced.
- The upper half may be notched to one-half depth without reinforcing in a nonbearing partition if only two studs in a row are notched and there are at least two unnotched studs left.

Attic joists can be crossed over, and first-floor joists can be crossed under, without difficulty. However, joists between floors must be notched if pipes are to cross them. For this reason it is much less difficult to run pipes between (rather than across) joists wherever

possible. If necessary, notch the joists as shown, and follow these rules:

- Never make any notch (at top or bottom) in the center half of a joist. Make the notching at the ends.
- Never notch deeper than one-quarter the height of the joist, and always reinforce it with a steel strap or 2 by 2 board across the cut.
- A hole, instead of a notch, can be made anywhere in a joist providing: (A) it is centered between the top and bottom edge, and (B) its diameter does not exceed one-quarter the height of the joist.

Notching can sometimes be minimized by cutting through flooring, subflooring, and stripping (if used), instead of the joists.

Toilet drains, which are a special problem, should be run between joists or below them whenever possible. But if necessary, one joist (only) can be cut off at its end (nowhere else) provided the cut joist is then securely anchored by a header to joists at each side.

Planning Bathroom Wall Alterations Assuming you have sufficient wall thickness to conceal the pipes, typical wall alterations are as shown. All fixtures are placed against one wall, together with a medicine cabinet. This is the simplest and most economical arrangement. Only the one wall (as shown) has all the finish stripped off. Other fixture arrangements will necessitate stripping of other walls also. The dimensions a, b, and c will be given on your installation plan. If you should decide to determine these yourself, however, they—and all other dimensions indicated by the letters—can be determined from the roughing-in dimensions furnished with your fixtures.

Planning Other Installations The installation of a sink is similar to that of a lavatory, and the kitchen wall is prepared accordingly. Walls through which a soil stack passes may also have to be stripped. If hubbed cast-iron pipe is used, a 6- to 8-inch-wide opening is required from top to bottom of the stack. If plastic, copper, or no-hub iron is used, however, lengths can be joined and pushed up inside from the basement to minimize the stripping required.

Whenever you can, locate the water heater as close as possible to the hot-water faucets it will serve. There

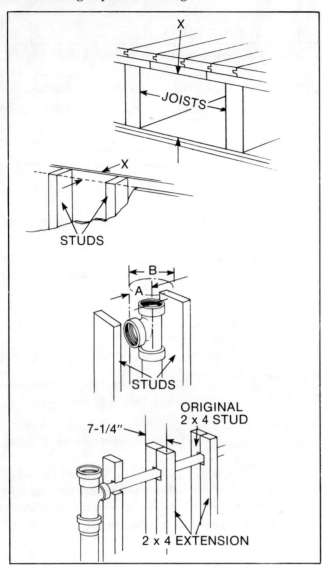

Rules for clearance of pipes: (top) necessary air; (center) turning diameter. B space is needed with all threaded pipes for the assembly of fittings. A is the radius of the turn; (bottom) method of building out wall to take the drainage pipe.

will be far less heat loss and hot water will "come on" quicker.

Wherever a trap or valve must be installed inside a wall or floor, provide a removable service panel so that it can be reached if necessary. Bathtub pipes are always hidden in the wall behind the tub. Plan the tub installation to provide such a panel opening into a room or closet behind the tub.

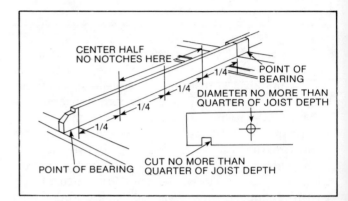

Rules for notching a joist.

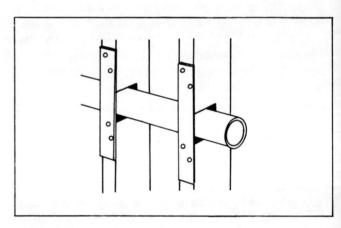

Reinforcing a notched stud.

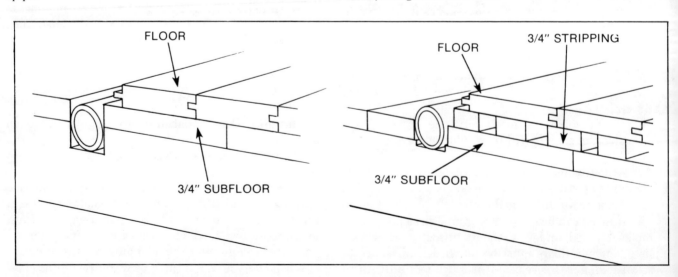

Other methods of running drain assemblies.

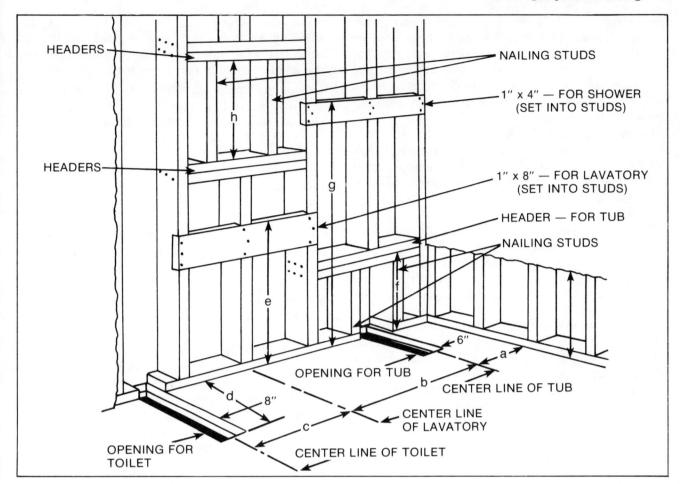

HEADERS

HEADERS

NAILING STUDS

1" x 4" — FOR SHOWER
(SET INTO STUDS)

1" x 8" — FOR LAVATORY
(SET INTO STUDS)

HEADER — FOR TUB

NAILING STUDS

6"

a

OPENING FOR TUB

b

CENTER LINE OF TUB

d

8"

CENTER LINE
OF LAVATORY

c

OPENING FOR
TOILET

CENTER LINE OF TOILET

Dimensions necessary for bathroom wall alteration: a = half the width of the tub. b = distance from the middle of the tub to the middle of the lavatory. c = distance from the middle of the lavatory to the middle of the toilet. d = distance from the finished wall to the center of the toilet-bowl outlet (usually 12 inches) plus 4 inches. When measuring from the face of the stud (not the finished wall), allow for the thickness of the wall finish (¾ inch for lath and plaster, ½ inch for sheetrock or ½-inch plywood, 1 inch for rock lath and plaster). Example: rough-in is 12 inches, and wall finish will be ½ inch; then d = 12 + ½ + 4 = 16½ inches. If the partition or wall must be thickened for pipe clearance, allow for thickening when measuring. e = distance from the floor to the top of the lavatory plus 2 inches (usually 33 inches). f = distance from the floor to the top of the tub. g = height of the shower (usually 5 feet). h = area needed to frame the medicine cabinet.

Plumbing Tools

Of all the "trades" a homeowner learns over the years of doing maintenance, repair, and remodeling work on a home, plumbing requires the least number of specialized expensive tools and equipment. For example, simple repairs like fixing a leaky faucet can be done with a screwdriver, an adjustable wrench, and a selection of washers. A plumber's helper, also known as a force cup, unclogs drains. If you do plumbing work beyond simple repairs, you will need some special tools, depending on the extent of the job and the type of pipe in the house.

Wrenches are of two general types, fixed and adjustable. Fixed wrenches have but one size jaw opening, while adjustable wrenches open or close to fit nuts and bolts of several sizes. Although adjustable wrenches range from 4 inches to 2 feet long, the average homeowner will find that 10-inch and 12-inch will handle most minor repair jobs.

Pipe Wrenches or Stillson Wrenches They are usually used to grasp pipes and other curved surfaces. Pipe wrenches in particular have solid housings and hardened steel jaws which provide excellent bite and grip. The primary difference between a pipe and a Stillson is that a Stillson wrench, with separate housing containing an adjusting nut, is subject to distortion and warping. Twelve-inch and 16-inch Stillson wrenches are most frequently used.

Chain Pipe Wrenches Costing a bit more than an equivalent-sized Stillson wrench, they offer easy handling in extra close quarters. This wrench consists of a forged steel handle to which is attached a length of heavy sprocket chain. The tool is used by wrapping the chain around a length of pipe and engaging the sprockets in notches on the back of the handle. Teeth on the face of the handle bite into the pipe while the chain grips it snugly to hold the pipe against the teeth and prevent slipping. This wrench turns the pipe in either direction and can be used like a ratchet wrench. That is, the handle can be loosened, shifted, and turned again without having to remove the chain from around the pipe. It can be used on round, square, or irregular shapes without crushing.

Lever Wrench or Locking Pliers This is built like a pair of pliers but serves as a wrench. It has compound lever action which enables it to be adjusted to size and then locked shut with a powerful grip. These tools are available with either straight or curved jaws.

There are also several other common plumbing tools used by the average homeowner.

Hex Wrenches They are designed for smooth surfaces such as chrome or highly-finished fittings and provide multi-sided, non-slip grip on any hex or square nut connection.

End Wrenches They are used where pipes are close together, close to walls, or against flat surfaces. It has a solid housing which prevents it from breaking or warping under normal use.

Strap Wrenches These are recommended for working with brass, aluminum, lead, soft metal, or plastic pipe because they grip pipe without teeth and do not damage the surface. A fabric strap, attached to a loop ring which is fastened in the curved head of a straight forged bar or handle, is pulled around the pipe, back through the loop, and over the head so that when the wrench is pulled tight, the strap grips the pipe.

Basin Wrenches These come in two types. One has fixed jaws opening at right angles to the shaft handle and is used in removing supply nuts and hose coupling nuts on faucet spray attachments under work tables, sinks, and lavatories. The other has spring tension pipe gripping jaws that are reversible by a flipover on the end of the drive shaft handle. It will grip nipples, odd-sized supply nuts, and jam nuts in hard-to-reach spots.

Nipple Wrenches or Extractors Extractors expand inside nipples where pipe wrenches cannot reach and are hex shafted for easy gripping.

Expanding Jawed Pliers They are more commonly called water pump pliers. Larger sizes can be used as a quick opening wrench, for loosening sink strainer jam nuts, and gripping flush valve jam nuts. All sizes are excellent for bench work, with or without vises.

Seat Wrench The seat wrench is the basic tool for plumbing jobs. It has several sizes of square and hex ends to remove faucet seats.

Seat Dressers Inexpensive types, often with ½-inch and ⅝-inch cutters, move in large quantities in most hardware stores. Better reseating tools have tapping attachments for reseating faucets with faulty and battered seat threads.

Handle Pullers They operate on the principle of wheel pullers and will remove corroded handles without scarring the chrome. Prior application of penetrating oil to the part is recommended.

Packing Nut Socket Wrenches They are available in sets and fit nearly all tub, tub and shower, and shower valves. They are hex on either end and hollow core to fit over faucet handles. Their importance lies in the fact that all faucet valve packing nuts and stem assemblies are brass; if the workman uses an open end or adjustable wrench, any real pressure will warp or break the nut or thimble. These are practically impossible to replace, and the cost of tearing out the wall—and tile—can be prohibitive.

Pipe Vises When you want to cut, thread, or ream pipe, you will need a pipe vise. There are two types available: yoke and chain. Both have specially designed jaws or chains for gripping pipe. A yoke vise will hold pipe ⅛ inch to 6 inches; a chain vise, pipe ⅛ inch to 8 inches.

A yoke vise has V-shaped jaws which grip pipe from above and below. The lower jaw is fixed; the upper jaw is raised or lowered by a screw. The pipe is held in the inverted V-shaped yoke which unlatches on one side and tilts so the pipe can be placed into it.

A chain vise is smaller with a fixed lower V-shaped jaw with teeth on which the pipe is laid and a bicycle-type chain fastened to one end. When pipe is inserted, chain is placed over it and locked in a slot on the opposite side.

Reamers Whenever a pipe is cut, both the inside and outside edges retain burrs. To remove burrs from the outside of the pipe, use a flat file. Burrs on the inside are removed by reamers.

Fluted reamers have straight cutting edges while spiral reamers have spiral shaped cutting edges.

Spiral reamers cut more easily, save time and work, and are often used by sheet metal workers to enlarge holes in sheet metal, conduit box outlets, etc., as well as smoothing inside edges of pipe.

Reamers are cone-shaped, with ratchet handles. Cutting edges can be sharpened, but this is difficult and time-consuming, and the small replacement cost usually makes it impractical.

Pipe and Tubing Cutters To make a clean, straight cut, use a pipe cutter. Most cutters have a single cutting wheel and two rollers which make smooth right-angle cuts. Such cutters are used to cut pipe ⅛ inch to 6 inches in diameter. Cutters are sized for pipe ⅛ inch to 2 inches, 1 inch to 3 inches, 2 inches to 4 inches, or 4 inches to 6 inches in diameter.

Major points of wear are the rollers, wheels, and pins on which they are mounted. When cutter wheels are worn out, they should be replaced; do-it-yourself sharpening is not advisable. If you plan to cut brass, copper, aluminum, or thinwall conduit tubing, use regular tubing cutters. Tubing cutters are similar to pipe cutters in that they have cutter wheels and rollers.

Some have a triangular blade-type reamer that folds out of the way when not in use. Tubing cutters and separate cutter wheels for plastic are available. Tubing cutters are sized to cut material with outside diameters ranging from ⅛ inch through 4½ inches.

Pipe Threaders Pipe to be threaded is held either in a pipe vise for hand threading or in the jaws of a chuck for threading on power equipment. When threading up to 2-inch pipe with power equipment, the die head or threader is stationary and the pipe revolves into the dies. With hand tools, the pipe is placed in a vise and the threader revolves around it. Thread-cutting oil must be used for best results.

Solder and Flux For home plumbers, solder is used to "sweat" copper fittings, forming a bonded joint between fitting and pipe. The proper solder is acid core unless electrical fittings are involved (in which case, use rosin core solder).

Flux (liquid or paste) and a small brush for applying the flux are also necessary. Used along with solder, flux prevents oxidation of metals as they are heated. It also chemically cleans the surface of items being soldered after they have been rubbed clean by steel wool or sanding cloth. By preventing oxidation, flux allows solder to flow freely, forming a good bond.

An acid core solder is available that is filled with scientifically prepared acid flux which is released in the proper proportions as the solder melts, thus doing two jobs at one time—ideal for the home do-it-yourselfer.

MAPP Gas To get a good sweat fitting, there should be no water in the pipeline, yet some homeowners find it difficult to thoroughly drain the line. This may cause a problem if the homeowner is using a propane torch which does not generate enough heat to dry out the line. MAPP gas, however, does reach a much higher heat (high enough for welding) that can dry out small amounts of water. Because of that much higher heat, it must be used carefully for it can melt the copper if held too long in one spot.

Pipe Joint Compound A big advantage of commercially prepared pipe joint compound is its ability to seal all joints, yet make disassembly, when necessary, easy by preventing seizure of parts by rust and corrosion. Pipe joint compounds come in tubes, cans, or drums.

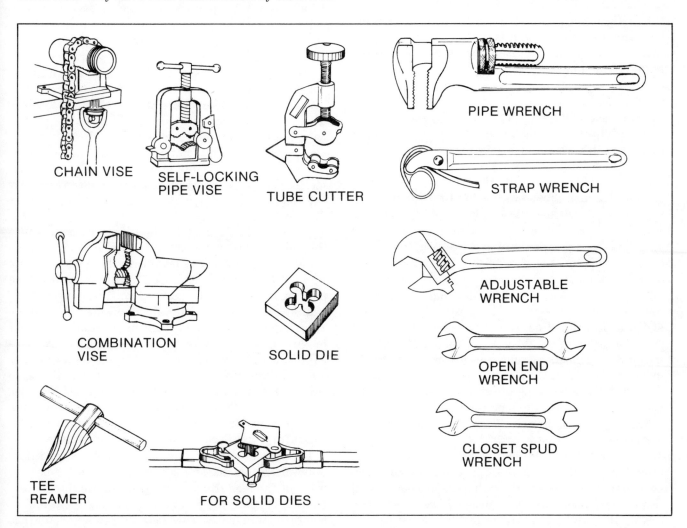

Common plumbing tools.

Assembling Pipe and Fittings

Standardized pipe fittings are available from any plumber's supply house and plumbing shop. These fittings are made for every type and size of job to be encountered in home plumbing. The term "fittings" includes all the connecting pieces that are necessary to join the lengths of pipe together to make up the plumbing system. Most of the fittings are threaded with standard pipe threads, except those used for soil pipes and drains, which have no threads.

One important requirement of good plumbing is to use as few of these fittings as possible. In order to keep them at a minimum and the pipe lengths as short as possible, as previously stated, you should make a thorough study of the plumbing system to be used and carefully make a detailed drawing of it.

Cast-Iron Soil Pipe

A cast-iron soil pipe is the most difficult for the home handyman to handle, but it is the least expensive. The tools necessary for the assembly job can usually be rented from your plumbing suppliers.

Measuring When pieces of pipe under 5 feet are needed, use double-hub pipe; then, when the pipe is cut, each end will have a hub. Carefully measure the length of pipe needed, making sure you allow the additional length necessary for engagement with the hub of the adjoining piece. Mark around the pipe with chalk where it is to be cut. Hub allowances are 2½ inches for 2-inch pipe; 3 inches for 4-inch pipe.

Cutting To cut service-weight pipe, use a hacksaw and make a ¹⁄₁₆-inch cut all around the pipe. Make sure your cut is square with the pipe to insure a clean, even break. Then, tap the pipe with a hammer until it breaks at the cut.

To cut extra-heavy pipe, first file around the pipe on the mark (using a triangle file) to scratch the surface and provide a guideline. Then, lay a piece of 2 by 4 flat on the floor and place the pipe across it. Take a hammer and cold chisel, use your knee to turn the pipe, and make a light cut all around the pipe. Continue cutting around the pipe, striking the chisel harder each time you go around, until the pipe breaks off.

Making Connections When the pipe is erected vertically, each piece is positioned with the hub end up. First make sure the ends of the pipes to be joined are clean and dry. Place the spigot (plain) end of the next-higher section into the hub to its full depth, and secure the upper pipe in position. Make absolutely certain the two lengths are perfectly straight up and down. Check with a level or taut cord. Then, light a

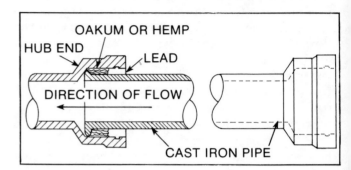

Bell-and-spigot joint in cast-iron pipe.

plumber's furnace and place the caulking lead in the melting pot. If your plumbing code requires a full inch of lead, use 1 pound of lead for each inch of pipe diameter. If it permits ¾ inch of lead, use 3 pounds for 4-inch pipe or 1½ pounds for 2-inch pipe.

Pack the joint with oakum while the lead is melting. Oakum comes in the form of a rope. About 1 ounce is required for each inch of pipe diameter. Wrap the oakum around the pipe at the joint, and drive it to the bottom of the hub space with the yarning or caulking iron. Continue packing until the joint is filled to within about ¾ inch from the top of the hub. Make sure the oakum is packed tightly and evenly, as this makes the watertight seal. The joint is now ready for leading. Warning: Heat the ladle by placing it alongside the melting pot before dipping it into the molten lead. An explosion may result if a cold or wet ladle is dipped into molten lead.

When the lead is molten, dip it out and pour it into the hub. Pour it evenly around the joint, and continue pouring until the lead is even with the top of the hub. Enough lead should be dipped to fill the joint in one pouring. When it cools, the lead must be packed down (caulked) to make the joint air- and watertight. Make sure the pipe is completely solid before starting. Other lengths of pipe may be positioned and packed with oakum while you are waiting. Then, use a hammer and caulking iron to force the lead down. An inside caulking iron is used to pack the lead against the pipe, an outside caulking iron to pack it against the hub. Tamp the lead firmly all around several times in order to obtain a tight seal between hub and spigot.

The procedure for caulking a horizontal joint differs only in that an asbestos joint runner must be used to prevent the lead from running out of the hub as it is poured. Prepare the pipes and pack in the oakum, as before. Now place the asbestos joint runner around the pipe, fitting it just above the hub and as tightly as

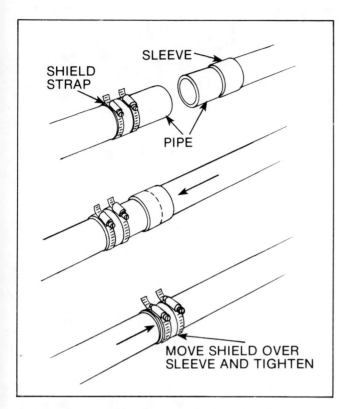

Steps in the assembly of a vertical cast-iron soil pipe.

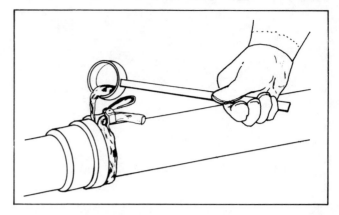

When pouring a horizontal leaded joint, an asbestos joint runner is used to keep the lead from running out of the hub.

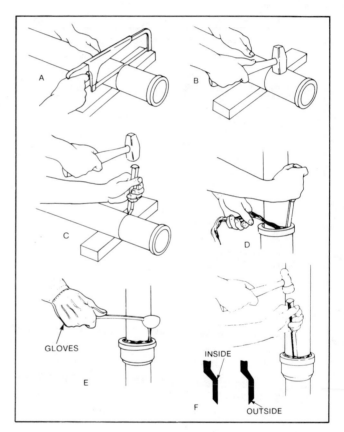

Method of joining no-hub iron pipe.

possible. The clamp should be placed at the top of the pipe to form a funnel for pouring the lead. Tap the runner down against the top of the hub to prevent the lead from running out. Then, take a full ladle of lead so that the joint may be made at one pouring, if possible. Pour the lead in at the top of the joint to overflowing. Remove the joint runner after the lead has cooled and solidifed. A good horizontal joint will always have an excess of lead showing above the hub. Cut off the surplus lead with a hammer and cold chisel. Caulk the lead into the hub, following the same procedure given before.

No-Hub Iron Pipe To join no-hub pipe, first cut the pieces to length with a pipe cutter, a hacksaw, or a cold chisel. Next, slide the neoprene sleeve onto the end of one pipe. Try inserting the other pipe to make sure ends butt tightly against the separator ring that juts out on the inside of the sleeve. The steel sleeve and clamp unit slips over the joint. Tighten the clamps with a screwdriver.

Steel Pipe

All fittings for galvanized steel pipe are threaded. The pipe is cut to the length required, threaded on both ends, and screwed tightly into the fittings. In many cases, if your order calls for pipe precut to exact lengths needed, these pipes will all be threaded by the supplier.

Measuring Pipe lengths to be cut should be measured very carefully, as allowance must be made for the threads needed to engage the fittings. The simplest way to measure is to use the face-to-face method. First measure the exact distance from face-to-face of the fittings. Next, refer to the allowance table to determine the extra length necessary for screwing into the fittings. Remember that double this length is necessary for two ends.

For example, the face-to-face measurement is 5 feet. If ¾-inch pipe is being used, the table below shows that ½ inch is needed for engagement with the fitting at one end. If both ends are to be engaged with a fitting, then twice this, or 1 inch of extra length, is required. The total length of the pipe will be 5 feet, 1 inch.

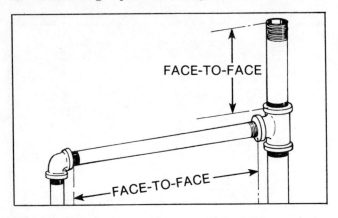

To get the correct length of pipe, use the face-to-face measurement method.

ALLOWANCES FOR STANDARD FITTINGS		
	Distance pipe is screwed into fitting (inches)	
Pipe size (inches)	Standard ("Ordinary") fitting	Drainage ("Sanitary") fitting
½	½	
¾	½	
1	⅝	
1¼	⅝	⅝
1½	⅝	⅝
2	¾	⅝
3		⅞
4		1

The illustration at upper right shows another accurate method of measuring pipe runs that is frequently used for steel pipe as well as for plastic. To measure pipe size by the so-called center-to-center method, proceed as follows:

1. Take the center-to-center distance for the pipe to be installed (dimension A).
2. Deduct the dimension shown at B, and then add the amount of pipe that enters the fitting, using the allowance for standard fittings table. (Again, you double this distance if the pipe goes into the fitting at two ends.)

Cutting Use either a hacksaw or a pipe cutter and cut squarely across the pipe. If the pipe is not cut squarely across and cleanly, threading will be difficult. For this reason, the pipe should be held in a pipe vise. Mount the vise solidly. Place it so there is ample room on each side for handling the longest pipe to be cut or threaded.

When using a pipe cutter, loosen its cutter wheel by turning the handle until the cutter will slide over the pipe (A). Place the cutting wheel exactly on the cutting mark and tighten the handle until the cutting wheel is forced slightly into the pipe. Apply thread-cutting oil to the cutter wheel and the pipe. Rotate the cutter one complete turn around the pipe. Tighten the cutter wheel and go around the pipe again. Repeat until the

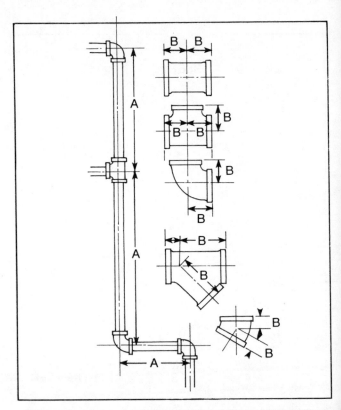

Another method of measuring pipe runs. This method is frequently used when measuring plastic pipe and fittings because fittings vary in size from manufacturer to manufacturer.

pipe is cut off. A hacksaw can also be used to cut the pipe to the desired length (B). Remove the burrs with a pipe reamer (C) or file.

When using a hacksaw, mark the pipe where it is to be cut and tighten it in place in the vise. Hold the saw at a 90-degree angle to the pipe, and make your cut with smooth, even strokes. With the pipe still in the vise, remove the burrs with a pipe reamer or round file.

Threading Care is required to insure clean-cut threads for engagement with the fittings. A stock and die are used. The stock contains a receptacle on one side in which the die sets, and an opening on the other side for inserting a guide. The guide makes it possible to start the die squarely. Each die is marked with its size. Select the same size die as the size of the pipe to be threaded. Loosen the thumb nut on the stock, slide the cover plate over, and insert the die. Make certain the printing on the die faces up toward the cover. Slide the cover plate back in place and tighten the thumb nut. Either an adjustable guide or individual guides can be used. Each individual guide is marked for the size of the pipe it fits. Select the correct guide, insert it in the opening in the stock, and tighten it in place with the lock bolt.

Place the pipe in the vise, and slide the stock over the end of the pipe with the guide on the inside. Push it onto the pipe until the die catches the pipe (D). Turn the stock slowly in a clockwise direction, keeping the die pressed firmly against the pipe. After cutting just

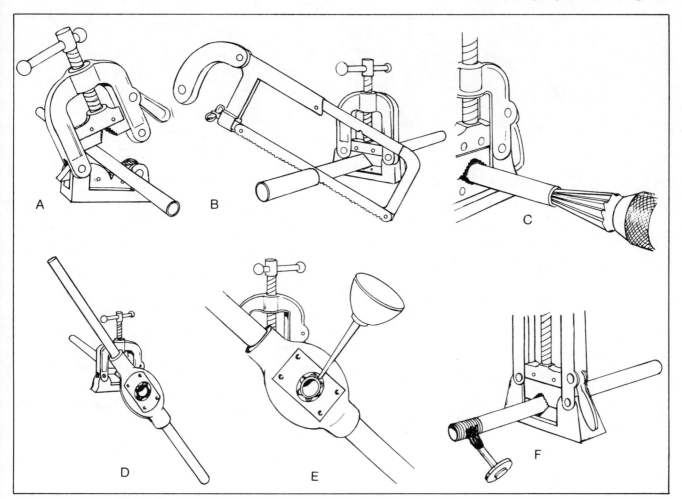

Steps in cutting and threading galvanized pipes.

enough thread so the die is firmly on the pipe, apply plenty of cutting oil to the threads of the die and the pipe end. Continue to turn the stock, backing off about one-quarter turn after each one-half turn forward, to clear away the chips. Continue threading, applying cutting oil often, until the pipe protrudes to the face of the die (E). To remove the tool after threading, turn it counterclockwise. Wipe off the surplus oil, and all chips from the thread, before using the pipe. Thread both ends, if necessary; then, before attaching the fitting, apply compound or special tape to insure a water-tight connection (F).

Making Connections Pipe wrenches are used for connecting this pipe and fittings. Use a 10-inch wrench for pipes up to 1 inch; an 18-inch wrench for pipes up to 2 inches; and a 24-inch wrench for pipes up to 2½ inches. (Wrenches are sized according to the overall length of the wrench.) Use an open-end, adjustable-head, or monkey wrench on nuts, unions, and valves, and to hold fittings with flat surfaces to be gripped. Do not tighten the jaws of a pipe wrench too tightly or they will tend to crush the pipe (this wrench gets tighter when you turn it). Actually, pipe wrenches are intended for turning pipes and other round objects, and not nuts, bolts, or flanges that have flat gripping

surfaces. The only exception is that a large pipe wrench may be used on a nut, bolt, or flange that is 1 inch or more across. Never use a pipe wrench on plated or polished surfaces as the finish will be marred. When using an adjustable wrench of any kind, always turn it so that the handle moves toward the open side of the jaws. This direction of turning tightens the wrench grip; opposite turning would loosen the grip and allow the wrench to slip. Never twist a wrench sideways. Use two wrenches when both fitting and pipe must be held. Position a left-hand wrench in front to loosen a joint, a right-hand wrench in front to tighten one.

Rigid Copper Tubing

Rigid copper tubing can be installed in much the same manner as wrought iron pipe, except that the ends are not usually threaded. This tubing is soldered to form a joint. Where long lines are exposed, rigid copper has the best appearance. The cutter used can be a hacksaw or a pipe cutter. If a pipe cutter is used, then you should have a reamer also, to ream out the burr that is always left on the inside of the pipe after cutting. If left in the pipe, the burr will cause a turbulence in the water and cut down the rate of flow considerably.

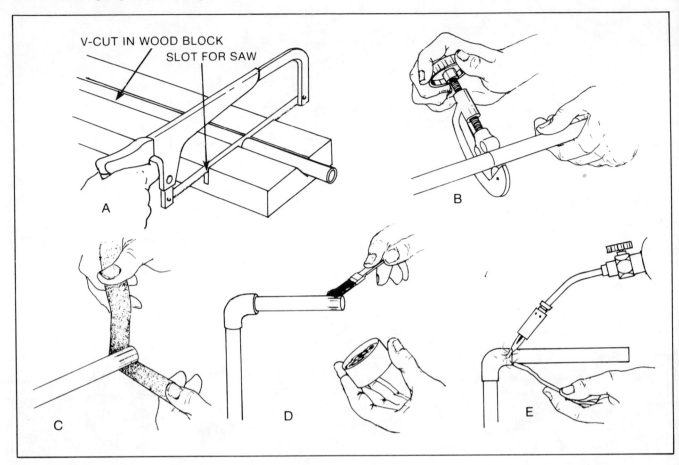

Working with rigid copper tubing.

Measuring This tubing is measured face-to-face in the same manner as galvanized steel pipe; but add on the depths of the soldering hubs in the fittings to be used.

Cutting If considerable tubing is to be cut, you will find it advantageous to build a jig with a V-grooved wood block and a slot for the saw cut at precisely 90 degrees (A). The jig holds the tubing and guides the hacksaw. This will enable you to get an even, square cut. Use a fine-tooth hacksaw blade (preferably a No. 24). You can also use a tube cutter (B) without a jig. After the tubing is cut, remove all burrs by reaming.

Making Connections Since clean surfaces are essential, clean and brighten the end of the tube and the inside of the fittings to be soldered with steel wool or fine emery cloth (C). Do not use a file as it will score the surface. Also, the tube end must be perfectly round, not out-of-round or dented. Apply a thin coat of noncorrosive flux or soldering paste (D)on the cleaned portions of both tube and fitting. Place the tube in the fitting and rotate it a few times to spread the flux coat evenly. Remove the excess flux from the outside of the fitting.

Heat the connection evenly with a propane or MAPP torch by applying the flame directly to the fitting. Use two torches simultaneously on the larger-size pipes, applying one to each side of the fitting to obtain uniform heating. When the flux bubbles out, remove the torch and touch the end of your solder stick to the edge of the fitting (E). If the fitting and tube are hot enough, the solder will flow and fill the joint immediately. (The solder will be drawn into the space between the tube and fitting, even upward into a fitting, by capillary attraction.) When a line of solder shows completely around the joint, the connection is filled. Do not hold the flame on the connection after it is filled as further heating will only result in a loss of solder, which might make it necessary to start over again. When the joint is completed, remove all surplus solder. If you make other solder connections to the same fitting, wrap the finished joint with wet rags to prevent the solder from melting. But, in any case, make sure the pipe and fitting do not move while the solder is cooling. Movement may result in a weak joint. If you have not overheated the connection, the solder will be firm in less than a minute. If a soldered joint should leak, the entire run of pipe must be thoroughly drained, and the fitting must be removed and cleaned before the joint is resoldered.

If you find it necessary to unsolder a connection, simply heat it until the solder runs, then pull the tube out. Use wet rags (as before) to keep from unsoldering other connections to the same fitting.

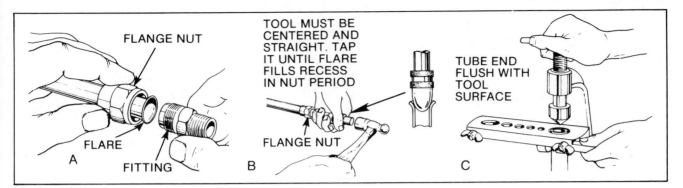

Methods of making connections with flexible copper tubing.

Flexible Copper Tubing

Like rigid pipe, flexible copper tubing can be installed with soldered joints, or the joints can be made with flare fittings. Flares are not recommended for inside wall construction, however, since it is possible for them to vibrate loose and drip. In open areas such as basements, garages, and utility rooms, flare fittings are fine. If used with solder-type fittings, flexible tubing is handled exactly like rigid tubing. The following describes the installation of flare-type fittings:

Measuring and Cutting Procedures for measuring and cutting flexible tubing are the same as those used for the rigid type. If the end becomes slightly out-of-round, flaring will reround it. After cutting, file off any burrs; the end must be perfectly smooth.

Making Connections Flare-type fittings have threaded flange nuts to hold the tubing in place. Remove the flange nut from the fitting and slide it onto the tube before flaring the tube. Flaring can be done either with a flaring tool or with a flanging tool. Put a few drops of oil on the tube end when flaring it. Hold the tube in your hand; do not rest the opposite end on a solid object (as it might be damaged). Slide the nut back up to the end of the tube (it cannot come off now), and start it on the fitting threads. Finish making the connection by tightening the nut securely with wrenches (preferably open-end or adjustable-head wrenches), one on the nut, the other on the fitting. This forces the flare at the end of the tube up against the cupped end of the fitting to form a tight seal. Never use one wrench alone; if the fitting is not held stationary by a second wrench, it will turn and loosen the connection at the other end. If a connection of this type must be broken, examine the flare to be sure it is still round and smooth. If damaged, the flare must be redone for reuse.

Compression-type fittings are even simpler to assemble and disassemble. A common use of the compression-type joint is the drain under a sink, although it can be used in other places. Slip a flange nut and a metal or rubber compression ring, in that order, onto the end of the tube. Push the end of the tube into the fitting as far as you can and tighten the nut down onto the fitting.

Plastic Pipe

Most major cities now approve the use of plastic pipe for earth-drainage systems and drain-waste-vent systems inside the home. This means that most plumbing, building, and supply houses carry plastic pipe and the material is increasing in popularity.

Measuring This pipe is measured face-to-face or in the same manner as galvanized steel pipe (A), but add on the depths the pipe will run into the fitting sockets.

Cutting Use a hacksaw or other handsaw, or even your power saw, to cut this pipe, but never use a rotary pipe cutter. Whatever saw you use, the blades should be fine-toothed (9 to 14 teeth per inch), with little or no set. To be sure the cuts are square when using a handsaw, you can build a jig to hold the pipe and guide the blade (B), or use a miter box. If you use a vise or other holding device, wrap the pipe in cloth or other protective material to prevent damage to the pipe surface. Ream the pipe with a standard reamer or with a pocket knife.

Making Connections When making a plastic-to-plastic pipe connection, clean both the fitting-socket surface and the surface of the pipe that will fit into the socket. With the proper solvent or cement for your type pipe (standard or drainage plastic piping) and a non-synthetic bristle brush, apply the solvent or cement generously to the inside surface of the fitting socket (C) and the outside surface of the pipe, over its circumference, covering an area at least equal to the socket depth (D). Press the pipe and fitting firmly together, turning the pipe one-quarter turn to evenly distribute the solvent (E). Next, hold the pieces together for about 15 seconds, or until "curing" has begun, so that the pipe does not push out from the fitting. Clean off any excess solvent.

After making the assembly, check immediately for the correct positioning of the pipe and fitting and make sure they do not move until the solvent weld has "set." Proper alignment should be determined before the

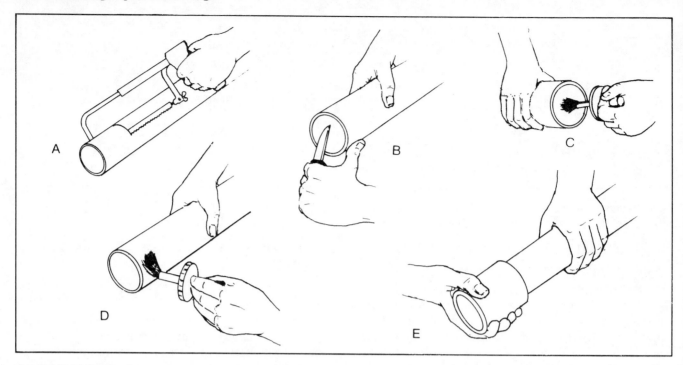

Assembling plastic pipe.

pieces are connected. If a solvent-welded joint has to be broken, the fitting must be sawed off and a new fitting used, unless you can salvage the fitting and a short length of pipe at each end. Do this by rejoining these assembled parts into a run with couplings.

To make a plastic-to-steel pipe connection, apply a special thread seal compound to both threads to join the plastic fitting to the steel. Afterwards, solvent-weld the fitting to the plastic pipe. Never use a pipe or chain wrench on plastic pipe or fittings. A strap wrench may be used or, on flat fitting surfaces, an appropriate toothless wrench. Where flange-type fittings or other types of bolted connections, such as expansion couplings, are used, a torque wrench should be used to tighten all bolts evenly. If the bolt tension is uneven, damage to the plastic components may result.

If you are able to break into your existing plumbing lines at a union, unthread the old piping back to the elbow you have chosen as a take-off point. Replace the elbow with a standard iron tee, which will then be threaded with a CPVC adapter. Be sure that the outlet of the tee points in the desired direction. Apply pipe dope to all threads and retighten the old iron piping into its former position. When tightening plastic-threaded fittings, use a strap wrench; pipe wrenches chew up plastic fittings.

If it is necessary to cut through a straight section of pipe, the procedure for installing a take-off differs. After cutting through the pipe in the area selected, remove the old pipe sections and thread in two CPVC male adapters with a coupling cemented to each. Make up a take-off assembly, consisting of two lengths of plastic pipe with a tee cemented between them. The total length of the assembly should be the distance between the coupling faces plus an allowance at each end to telescope into the coupling socket. Allow one diameter for each joint. Check the take-off assembly by holding it in position before cutting; then cement the joints, making sure the tee is pointed in the desired direction.

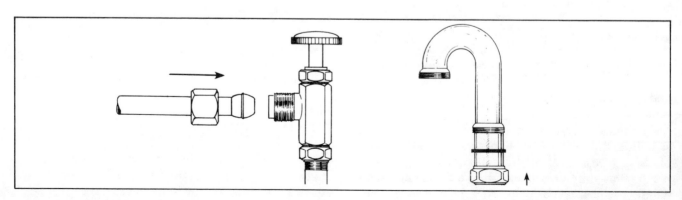

Assembling a typical compression-type fitting.

Installation of a Drainage System

When making any plumbing installation, begin with the drainage system. It is easier to install the water-supply piping afterwards. Study your plan to familiarize yourself with the arrangement of the pipes and fittings when installed. When installing any drainage system, you should follow the sequence of steps given on the following pages. However, each installation may vary in details from others, so the illustrations that follow should be considered to show only typical solutions.

Preparation and Positioning of Toilet Drains

Before installing any pipe, you must make all necessary wall and floor openings for the stack and the bathroom plumbing. If your bathroom is above a room ceiling, cut away an 8-inch-wide strip of flooring from the center line of the toilet-bowl outlet to the wall, then back through the wall finish and bottom partition header. This will provide space for installation of the stack T and closet-bend assembly from above. If there is no ceiling below, cut just two holes, one for the stack, the other for the closet-bend opening, and install the parts from below.

If you are using plastic drain pipe, the toilet-drain assembly will consist of a T, closet bend, two short lengths of pipe, and a floor flange. Copper pipe differs only in that there is no closet bend and a one-quarter bend is used instead. To make either type of assembly, first calculate the vertical length of pipe needed between the flange (positioned to rest on finished floor)

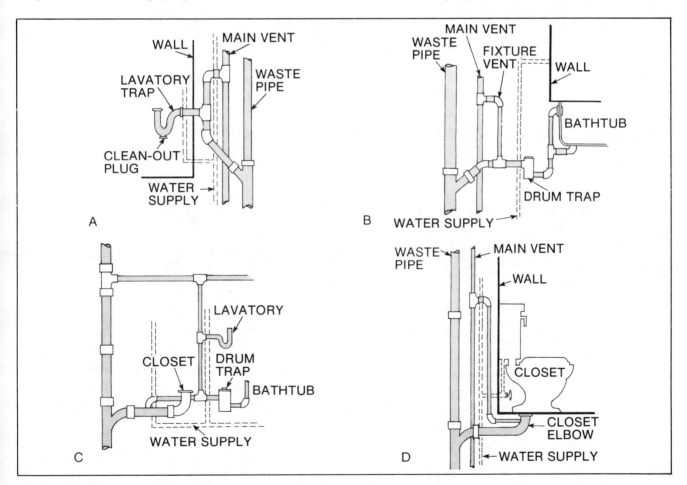

Examples of drainage piping in typical installations. The lavatory installation in A shows the drain line connected to the waste pipe at an angle. B shows a method of connecting a fixture vent to the main vent. In C, all three bathroom fixtures are connected to a common vent pipe. The toilet installation shown in D connects the vent to the main vent above the fixture.

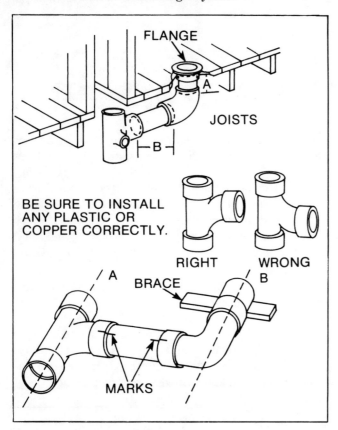

Toilet drain hung in place.

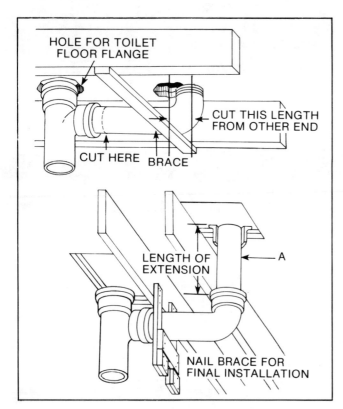

Assembly of toilet drain.

DWV systems may be of cast iron (both hub and hubless), plastic, or copper.

and the bend so that the latter will be at the depth required by your installation. Weld or solder the bend and vertical pipe lengths. Next, with the T and bend held in their places (any means will do for now), determine the length of horizontal pipe required. Loosely assemble the T, horizontal pipe, and bend, with the parts lying on the floor and braced so the center lines are parallel, and mark the pieces. Then weld or solder them together, being guided by the marks.

When cast-iron pipe is used, the fitting is much the same. If assembly will be between joists, only a T and closet bend are used, and the distances are adjusted by breaking off parts of the toilet bend (along scored lines). If assembly will be below joists, a T, hub top with a one-quarter bend, and vertical length of pipe are needed, and A and B are established as for plastic

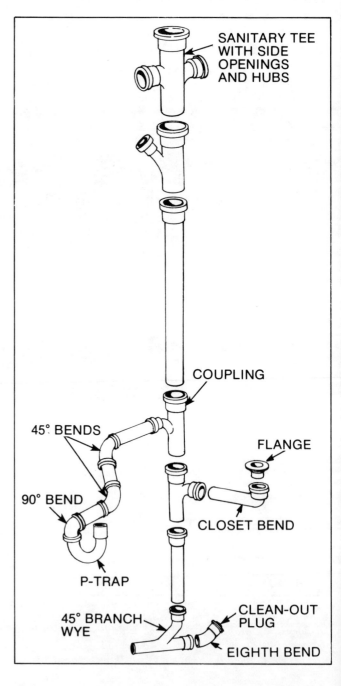

or copper. When all the pieces are ready, assemble and mark them as above, then brace them as required, to caulk the joints. After completing the toilet-drain assembly, brace it in position. Do not install the floor flange, however, until other work has been completed and you can finish the bathroom floor, so the flange will rest on it.

Installing Building Drains

There are two types of building drains: the underfloor drain and the suspended drain. For the former, if you have a plastic stack, either plastic (4-inch sewer type) or 4-inch cast iron may be used for an underground building drain. With either a copper or a cast-iron stack, 4-inch cast iron generally is used for this drain. When converting from one kind of pipe to another kind, from plastic or copper to cast iron for instance, you must use an adapter, generally located at the foot of the stack, just above ground (or basement floor).

If you have a plastic stack, this drain also may be 3-inch plastic; with a copper stack it may be 3-inch copper—up to the Y assembly located in the basement wall or floor. With a cast-iron stack, the drain will be 4-inch cast iron. In most cases, two cleanout assemblies using T-Ys generally are used, one at the foot of the stack, the other where the drain goes through wall or floor. The drain, from the second (in wall or floor) assembly outward, will be 4-inch plastic or 4-inch cast iron (the same as in an underfloor drain, discussed before). If only one cleanout assembly is used, all horizontal drains should be 4-inch pipe.

Locating the Foot of the Stack (Underfloor Drain) Use a string and plumb bob (or similar weight to hang string straight) and hang the string through the exact center of the toilet-drain T with the bob (or weight) ½ inch from the basement floor (or earth). When the string is straight and stationary, mark the floor (or earth) where the bob would touch.

Preparing the Building-Drain Trench Draw a line on the floor from below the bob mark to where the building drain will go out through the house foundation. Dig a 2-foot-wide trench with this line at its center, digging deep enough so that the bottom of the trench at the start (where mark was) will be approximately 1 foot below the surface of the finished basement floor. Now grade the trench bottom from this starting end downward to the other end, and firmly pack the bottom of this established grade. Proper grading is important. The grade should be 1¼ inches per 5 feet.

After completing the inside (above) trench, go outside and extend the trench at least 5 feet outward from the house. Connect the two trench portions by tunneling under (or through) the foundation. Make certain that the grade is correct all the way to the outside, and

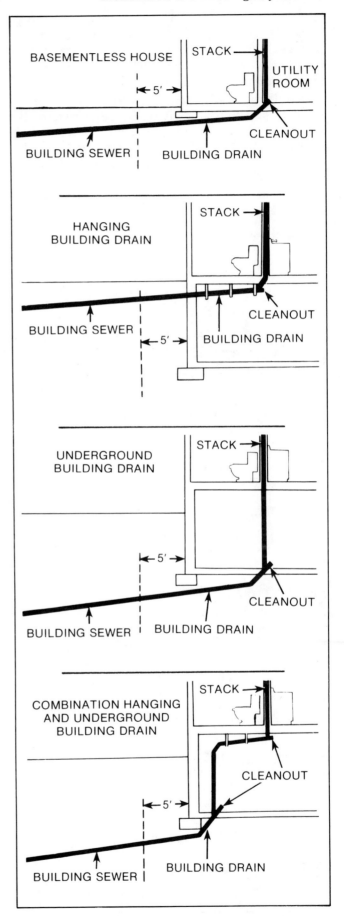

Drain arrangements for various house designs.

that the bottom is firmly packed. If you are installing a basement floor drain or other branch building drains (as from a secondary stack), prepare the trenches for these in the same way, each graded to slope down so that the branch can join the main building drain at a Y.

Installing an Underfloor Drain Assemble the cleanout assembly first. If this is to be plastic, the assembly may consist of a cleanout T and two one-eighth bends, or a Y with a cleanout adapter and plug and a one-eighth bend. If it is to be cast iron, there may be a test T and two one-eighth bends, or a Y with cleanout ferrule and plug and a one-eighth bend. In either case, use the same method of aligning the parts

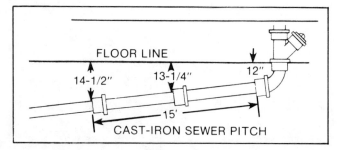

Proper pitch of a sewer line.

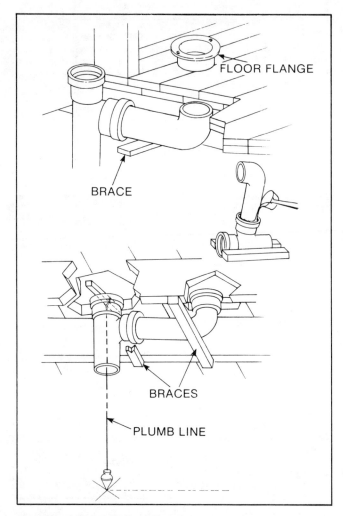

Installing an underfloor drain.

as for the toilet-drain assembly (discussed before) and weld or caulk all the parts together. Position the above assembly in your trench, centered under the (still hanging) plumb bob. Brace it carefully so that it is exactly centered and aligned with the plumb bob and string; then pour concrete under and around it to permanently cradle it in this position. When the concrete has set, install the remainder of the building drain, out to 5 feet beyond the house foundation, setting in any Ys needed for branch drains. It is a good idea to install all other underfloor drains now so that you can finish the basement floor and clean up before finishing the drainage installation. A floor drain assembly is prepared and cradled in concrete as described before.

Preparing and Installing Foot-of-Stack Cleanout Assembly Preassemble the plastic, copper, or cast-iron pipe in the same manner described for the cleanout assembly used with the underfloor drain. Weld, solder, cement, or caulk all joints, installing, if necessary, a reducer or adapter at the top of the assembly to convert it for connection to the stack. When ready, join this assembly to the toilet-drain assembly, either directly or with a length of pipe between. Length A may be needed if the drain must go through the wall at a certain height and can be determined by using a chalk line. Brace the assembly firmly in place. *Note:* If length A extends up to a second-floor bathroom, build the stack up from the cleanout assembly, then fit and install the last piece.

Installing a Suspended Drain If you have not done so, use a chalk line to determine where the drain

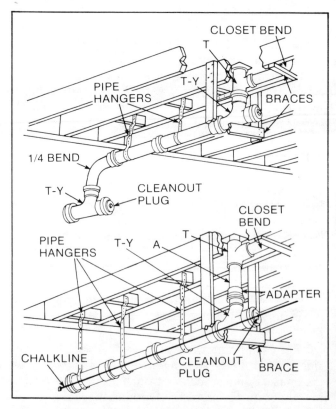

Preparing and installing a foot-of-stack cleanout assembly.

will go through the wall. Remember, it must pitch down 1¼ inches per 5 feet. Make a hole through the wall at the place indicated and prepare the trench outside for at least 5 feet out from the wall. Install the drain from the foot of the stack outward to 5 feet beyond the wall. If the drain includes a second clean-out assembly in the wall, preassemble the T-Y and one-quarter bend (with any straight length of pipe needed between them) and brace these in position to help you fit the run to them. Permanently brace each section of pipe (every 5 feet) to joists above as you install it, and remember to keep the drain properly pitched.

Completing the Main Stack

If you have a suspended building drain (see above) the stack is now complete up to the toilet-drain assembly. If you have an underfloor drain, this portion is yet to be done. Build it straight up from the cleanout assembly, fitting the last piece as shown. To install the last piece, temporarily raise the toilet-drain assembly.

Build the stack upward from the toilet-drain assembly and out through the roof. If there are any drainage or vent Ts needed, install each as you come to its place. Remember that each branch-drain T must be at the proper height so the drain will slope properly (1 inch per 4 feet). Refer to your plan on roughing-in measurements and plot each T position by using a stretched chalk-line to represent the drain that will be connected to it. Re-vent runs, on the other hand, slope slightly upward to the stack. With a cast-iron stack, special vent Ts are used; but with plastic or copper, sanitary Ts are used, and these must be inverted (up-side down) in the stack when used for vent connections.

If there is a bathroom above the first one, preassemble and brace in position the second toilet-drain assembly (in same manner as the first), before building the stack up to it. This will make it easier to accurately fit the stack pipe to place it at the correct height. Should a roof rafter or other obstruction require turning the stack, make an offset. One-eighth bends are used if offset occurs in a drainage portion of the stack; one-quarter bends if in a vent portion. The topmost length of the stack should extend at least 6 inches above the roof; however, it must also be high enough to prevent snow blockage. The opening in the roof through which the pipe passes should be made watertight with a roof flashing. The flashing is adjustable to roof pitch. Make sure the roof shingles (or other covering) overlap the flat part of the flashing at the top and sides. Seal the flashing to the stack pipe by peening in the sides against the pipe.

In very cold climates, the part of a vent above the roof should be at least 3 inches in diameter to prevent frost closure in cold weather. Where individual vents

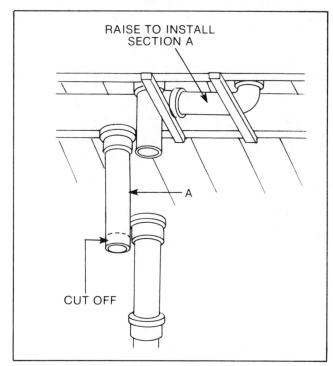

RAISE TO INSTALL
SECTION A

A

CUT OFF

Completing a suspended main stack.

are used for fixtures, 1½-inch pipe is recommended. Vent increasers may be used to increase the diameter of the vent stacks above the roof.

Installing Branch Drains and Re-Vent Lines

Each branch drain is installed outward from the stack to end at the location required by your plan or roughing-in measurements. These drain runs may be plastic, copper, or steel and generally come in 1½-inch sizes. The same is true of any re-vent lines required. With plastic or copper, sanitary Ts are used to join re-vent lines to the branch drains, and the re-vent lines are built upward from these Ts to join the inverted vent Ts already installed in the stack. With steel pipe, either a vertical vent T or a horizontal vent T is required, and the vent run is built out and down from the stack T so that the final joint can be made by caulking the (unthreaded) steel-pipe end into the hub of the drainage-line vent T. Two (or more) vent runs may be joined by a T (or Ts) to run as a single line into the stack.

Installing a Secondary Stack (if necessary) If a secondary stack contains a toilet, it is installed just like the main stack, and its foot-of-stack cleanout assembly must also be positioned and joined to the building drain in the manner already described. If there is no toilet, however, the cleanout assembly location is not quite so critical. When installing the building drain and the branch drain to it from the secondary stack, simply hang a plumb bob approximately where the secondary stack will be, and position the cleanout assembly accordingly. Afterwards, build the secondary stack straight up from this cleanout assembly and install branch drains, etc., following the same general procedures given for a main stack.

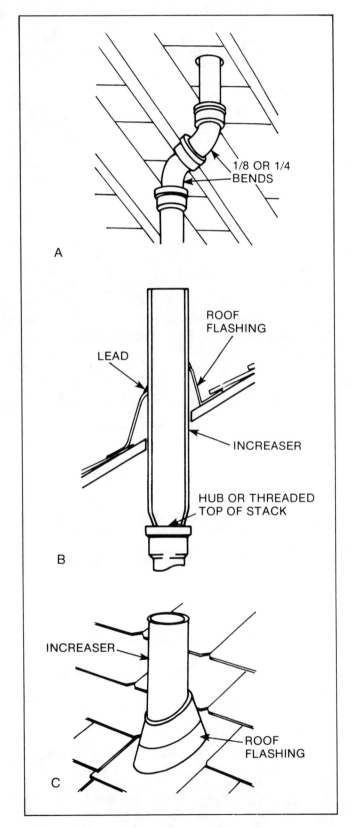

Roof work necessary for a main stack.

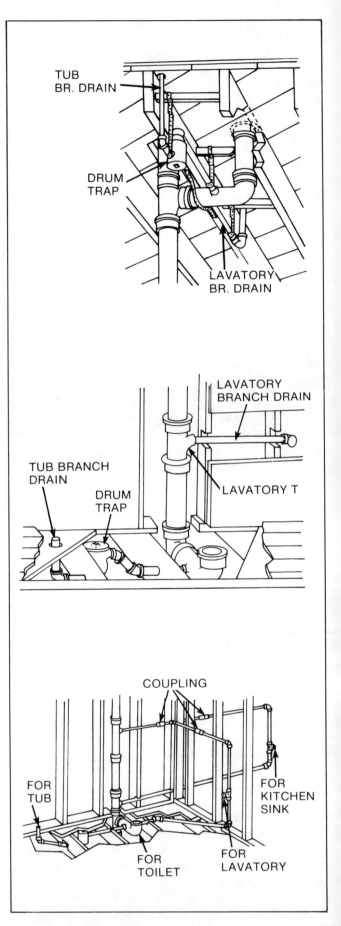

Techniques of installing branch drains.

Tapping into an Existing Stack

If you install a new fixture close to an existing stack, it is possible to tap into it, but you must check building codes first. If you have a plastic or copper stack, the job is much easier than with cast iron or steel. Usually, with a suspended stack made of copper or plastic, you can simply cut a piece of pipe or drill a hole to accept the fitting. On the other hand, steel or galvanized pipe must be supported and an entire section of pipe removed.

First run the drainage line from the fixture to a convenient space in the stack. If the stack is plastic, cut a hole in the stack where the pipe will enter. Make the hole a little larger than the opening in the T or Y fitting. Use a special fitting called a saddle, or slit a plastic sanitary T lengthwise and slip it over the stack. Solvent-weld it in place. Use a clamp to hold the joint.

If the stack is copper, cut out a piece about the size of the fitting, using solder slip couplings, if permitted by code, to seal the new joint. If the soil stack is cast iron, you might have to disconnect the fixtures above the new connection to be able to move the stack up and remove an entire section of pipe. Drill or chisel the joints loose. Then, replace the pipe section with a shorter piece of pipe plus the new fitting, which together measure the length of the previous section. Or, you may be able to use a no-hub fitting to tap into an existing stack. Install the cross at the desired height in the established stack.

Testing the System

After work is completed, and before it is concealed, the drainage system should be tested for leaks. Local plumbing codes usually include a standard testing procedure. Where no code is in effect, the drainage system may be tested as follows: (1) Tightly plug all openings, except the highest one; (2) fill the system with water, and let the water stand for about 15 minutes. Then check the entire system for leaks.

The system can be checked by sections. If done that way, test each section under a head (depth, measured vertically) of water of at least 10 feet to be sure that each section and joint will withstand at least that much pressure.

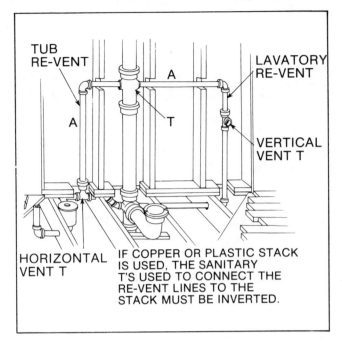

Installing re-vent lines.

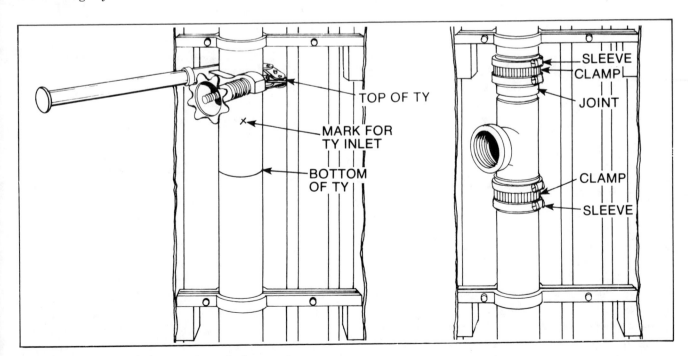

How to tap into an existing cast-iron stack.

Installation of a Water-Supply System

When compared with the installation of a drainage system, water-supply system installation is easy. The job can be divided into two parts: the house-service line and the internal lines.

The House-Service Line

This line is the outside line, from source to house. It should be laid in as straight a line as possible from the water source to the most convenient point at which it can enter the building. Also, it should be buried below the frost-line and deep enough so traffic cannot disturb it.

If the source is a city or county water main, the water authorities will probably want to install (or at least approve) the house-service line. This line will end at a water meter (if one is used) installed by the water company. It should be of ample size to furnish all the water you will need at times of peak demand. This size may depend upon normal pressure in the water main, especially if it is usually low, or if your requirements are above average.

Should you have a private water source, the house-service line is considered to be all the piping up to your pressure tank, from which the internal main supply lines lead to the various branch supply lines and fixtures. Whatever method is employed, the piping must run below the frost-line, from the well casing through the foundation wall. Check your plumbing code for specifications.

Internal Lines

Once the water passes through the water meter (if municipal water supply) or into the house from a private source, the service line becomes the internal portion of the water-supply system. In most cases, immediately after the meter or entrance into the house, there is a main shutoff valve (also called a stop-and-waste valve). Closing this valve turns off the water supply to the entire house. It is usually located at the lowest spot in the internal water-supply system. In most systems, immediately after the valve there is a drain valve which allows the water to be easily drained if the house is to be left unheated during freezing weather. Every member of the family should know the location of the main shutoff valve so the water can be turned off quickly in an emergency.

From the shutoff valve, the internal water supply divides into two branches. One of these branches continues to the fixtures, delivering them cold water. The other branch flows through the water heater, then to all the fixtures that use hot water. Pipes carrying cold water are called cold-water mains, while those carrying hot water are called hot-water mains. Branches of these mains that lead to fixtures are simply called fixture branches, hot and cold. Cold water should always be valved as it enters the water heater. This valve allows turning off the entire hot-water system from one point. If there is a water softener in the system, the water flows through it. Water bypasses the softener for outdoor use and sometimes for toilet flush water. Softened water then flows to the water heater.

When running risers (pipes) to fixtures, remember that both the hot- and cold-water lines are continuous throughout the house (as mentioned, the only place they join is at the water heater) and that the hot is always on the left, the cold on the right. All horizontal lines should have a slight slope back to the house-service line. A stop-and-waste valve should be installed at this low point so that the internal-supply system can be drained when necessary. Pipe sizes should be chosen so that each line will have a sufficient capacity to serve all the fixtures to which it leads. The supply lines should, of course, be large enough in diameter to deliver all the needed water when all fixtures are operating. The following chart gives gallons per minute (gpm) requirements for various household fixtures: bathtub 5—8, tank toilet 2—3, flush valve 30—40, laundry tub/sink 5, lavatory/shower 5, shower-water conserving 2, garden hose 5—10.

Provide shutoff (gate) valves wherever convenient and on both lines at the water heater. Also, when using threaded pipe, install union couplings wherever the line may later have to be broken, such as in the lines to the hot-water heater (close to it).

Usually the main supply lines are installed under the first-floor joists and are attached to these by hangers. The pipes can be run at right angles or parallel to the floor joists. Union L's and T's can be used to advantage when connecting branch lines to mains; they eliminate the need for careful measuring and fitting. The hot- and cold-water lines run parallel and should be spaced at least 6 inches apart (unless the hot-water pipes are insulated) to prevent the cold-water pipes from absorbing heat from them. Remember that the hot-water line is run into the left side of each fixture (as viewed by a user facing the fixture).

When laying out supply or feeder runs, make them as short and straight as possible. The longer the pipe and the smaller the diameter, the greater the loss of

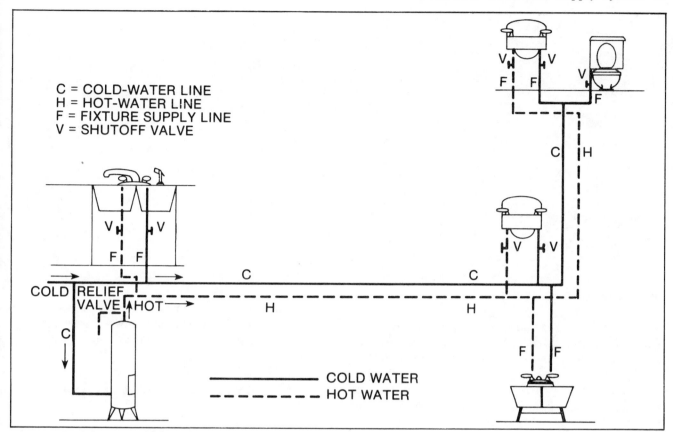

C = COLD-WATER LINE
H = HOT-WATER LINE
F = FIXTURE SUPPLY LINE
V = SHUTOFF VALVE

——————— COLD WATER
– – – – – – – HOT WATER

Parts of a typical water-supply system in a two-story house with a basement.

pressure. Pressure is also lost whenever water passes around bends and through elbows and other fittings. Also, try not to place any in outside walls, for no matter how well insulated, sooner or later they will cause trouble by freezing. All risers and feeder lines must be well blocked and securely fastened to support risers; use 2 by 4 braces with cleats nailed against a pipe on both sides with another nailed across the pipe to the cleats. When a riser runs from the basement to the first or second floor, a solid block of wood should be located under the elbow at the lower end to support the weight of the vertical pipe. It will also keep the riser from hanging on the branch lines to the fixtures.

Pipe hangers and clamps should be employed to support overhead runs whenever possible. In the basement, where the run may be across the floor joists, the pipe can be clamped up against the joists with pipe straps or clamps or can be supported just below the joists on pipe hangers or "plumber's tape," depending on the available headroom. (Plumber's tape is a long strip of galvanized steel about ¾ inch wide and perforated at intervals to receive ¼-inch bolts.) Secured by bolts to the structural members of the house, lengths of this tape form loops in which the pipes rest. All pipe— even steel and plastic—expands when heated. Leave space in the hanger for the pipe to expand. Very long copper or plastic runs also require a special expansion fitting or loop. Steel and rigid copper tubing should have supports every 7 to 10 feet. Soft copper tubing

needs support every 2 feet.

Cut notches or holes in joists or studs for pipes to pass through. If the pipe diameter is less than one-fourth the joist depth, holes are easier and quicker to make than notches. Holes must be 2 inches from either edge of the stud and not more than one-quarter the depth of the joist or stud. Notches also should not exceed one-quarter depth and can only be cut out of either side—never in the middle. If steel reinforcing straps are used, a 2 by 4 stud can have round or square notches cut out up to 2¼ inches wide.

Insulation When it is absolutely necessary to run supply lines in or near outer walls, it is most important to insulate the pipe even though the wall itself is insulated. In fact, all cold-water pipes throughout the house should be covered with insulation. Such pipe insulation is not only a safeguard against freezing, but it will prevent summer condensation with its subsequent dripping and water-staining of walls and ceilings or the water-rotting of studs and wood around the pipe. Condensation in some areas of the country can become so great as to cause an actual flow of water down the outside of the pipe. Placing insulation around hot-water lines conserves hot water and achieves a saving on heating bills. Several types of wrap-on insulation are available at plumbing supply shops and hardware stores. Install them as directed by the manufacturer.

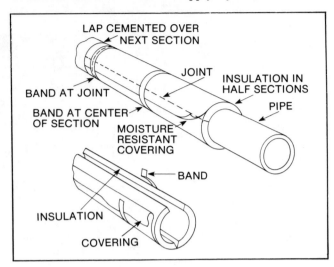

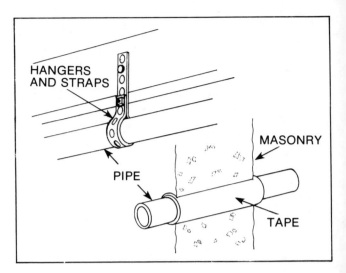

(A) Hanging pipe with a strap; (B) Using plumber's tape to pass through a masonry wall.

One method of applying insulation to a pipe.

Connections Through the Floor Branch lines for the hot- and cold-water main supply lines are run below the floor joists to a first-floor bathroom. Fasten the pipes to the joists with pipe straps. If the pipes cannot be run below the joists (as with a bathroom on the second floor), run them just below the floor and across the tops of the joists, notching each joist to recess the pipes. Use plugs and caps to stopper the branch-line ends while patching walls and floors.

Connections Through the Walls The illustration shows a typical installation in which the supply lines are run under the joists and the partition plate is notched for the vertical branch lines to the fixtures. If the bathroom is on the second floor, the supply lines are generally run up next to the stack and the branch lines are run horizontally through notches in the partition studs. The other illustration shows how a water supply from the floor and from the wall may be handled.

Provisions for a Hot-Water Heater Plan your supply line to include the requirements for your hot-

water heater. If you are remodeling your system, you will usually replace old supply lines with new, leaving the water heater in the same location.

Avoiding Noisy Pipes

Vibration can cause fittings in pipe joints to loosen and leak. The most common vibration-causing villain is water hammer, which occurs when water traveling under pressure in the pipes is suddenly cut off. The pressurized water slams against the closed valve and the shock undulates all along the water line. You may be able to hear it by turning the water on full blast then shutting it off quickly. The solution in the case of kitchen sinks, lavatories, and tubs is to simply shut the water off gently. Water hammer is a bigger problem with washing machines and dishwashers which have automatic valves that shut off abruptly.

The permanent solution to water hammer for these fixtures is an air chamber. These chambers are short, "dead-end" lengths of pipe in which air is trapped and compressed by the water flow in the line. This trapped air absorbs the shock caused by sudden starting or stopping of the water flow in the line (as when faucets are turned on or off) and thus prevents water hammer. Use a serviceable chamber in each branch line serving a hot or cold faucet. To provide air chambers when the pipes are in the floor, run a horizontal line from each supply line to the nearest partition. Then make a 24-inch air chamber inside the partition to serve all the faucets on each supply line.

Water Treatment

Once you have done the work to supply water to fixtures, you want to be sure it is safe to use. Most water supplies, although safe to drink and reasonably clear, are loaded with chemicals, minerals, gases, unpleasant tastes and smells, dirt, and decayed vege-

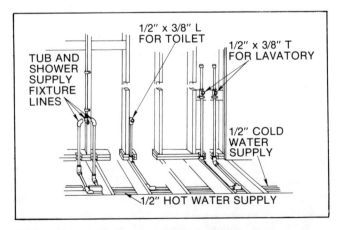

A typical installation in which the supply lines are run under the joists and the partition plate is notched for the vertical branch lines to the fixtures.

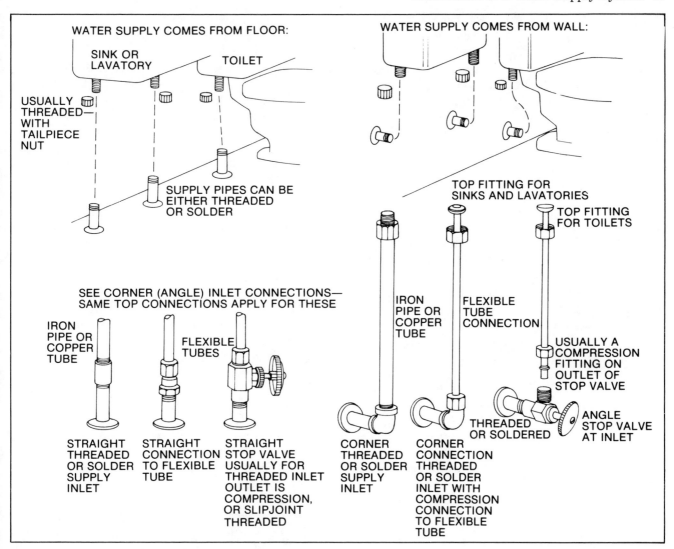

Methods of handling piping when it comes from the floor and from the wall. Most of the fittings shown are available with either threaded or solder inlets.

table matter. As the rain falls to earth it accumulates soot and dust and absorbs carbon dioxide. When the water filters through the soil it takes in more dirt, minerals, organic matter, and bacteria.

Various types of water purifying equipment are on the market today. First, however, you must determine what kind of contaminants, if any, your water contains.

Hard Water When excessive quantities of nontoxic minerals are present in the water supply, the life expectancy of the plumbing system can be cut to a serious degree. Certain minerals cause deposits to form in the pipes. Eventually, these deposits become large enough to interfere with the flow of water. In extreme cases, they have been known to stop it entirely. It is also likely that rust and corrosion may advance at a great rate owing to the presence of these minerals.

In another respect, minerals may not be as harmful as they are inconvenient. Probably the most common inconvenience that can be laid at the door of "hard water" (the high-mineral-content water we have been

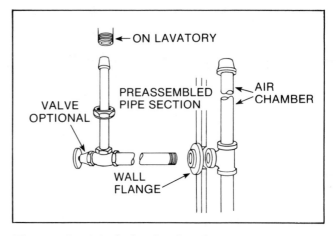

The use of a 12-inch chamber for a lavatory.

talking about) is the difficulty of washing with it. Soap will not lather and forms hard-to-dissolve curds, leaving an unsightly film on skin and clothes. To make hard water soft, the mineral content must be eliminated. Commercial liquid water softeners can be added to the clothes washer each time you wash. A permanent solution is to pipe the domestic water supply through a device called a water softener.

Most water softeners have few moving parts and consume little power. The water is treated as it flows through a special chemical that removes the objectionable minerals. Depending on the hardness of the water, the rate of consumption, and the unit's capacity, there comes a time when the chemical must be regenerated, or cleaned and renewed. Different types of chemicals and equipment may be required to treat your water problem. For this reason, a chemical analysis of your water should be made. This may be performed by various local agencies, or you can obtain a kit from your plumbing supplier. After the water has been analyzed, you can determine the water softener that is best for you.

To install a water softener, you will need to cut into the existing cold-water supply line ahead of your water heater and ahead of any cold-water outlet that you want softened. Connections are made to the softener and then back to the water line. All of the lines downstream of your softener will then carry softened water. A grounded, plug-in connection to your household electrical system also is necessary. In addition, you will need an adequate drain for the softener. In order to minimize plumbing, it is generally recommended that you locate the softener as close as possible to the place where you interrupt the line. However, remember it is much easier to extend plumbing lines than it is to relocate a drain or move a major installation, such as a furnace or water heater. Consider these factors when choosing your softener location:

- If you wish to conserve soft water, provide a separate hard-water line ahead of your softener to your outside faucets for lawn service.
- An approved 115/120-volt, 60-hertz grounded receptacle should be within reach of the 6-foot power cord.
- Allow sufficient space around the softener for easy access to add the salt and to make all connections.
- Softeners should not be exposed to freezing temperatures and must be installed and operated in an upright position.

The actual assembly of the softener and its complete installation should be made as directed by the manufacturer. Should the water pressure in the house line run in excess of 75 pounds, it is recommended that a pressure regulator be installed in the supply side of the softener. This not only insures longer life, but it also cuts down water noise throughout your house.

Other Water Problems Some water problems can be detected by the taste, smell, or appearance of the water as it comes out of the tap. For example, you might notice a rusty tinge to the water and rust-colored stains around the sink and bathtub drains, in toilets, and sometimes in washing machines and on clothes. The water may also look clear as it comes out of the faucet but become red when exposed to air. This condition is known as red water and indicates high iron content. It is not dangerous to drink, but it does not taste good; if not corrected, it will corrode pipes and cause pinhole leaks. A water softener will take care of a minor red water problem, but you might need a separate water filter if the problem is severe.

You can test for organic matter, such as algae, silt, mud, or sand, by taking two samples of your tap water,

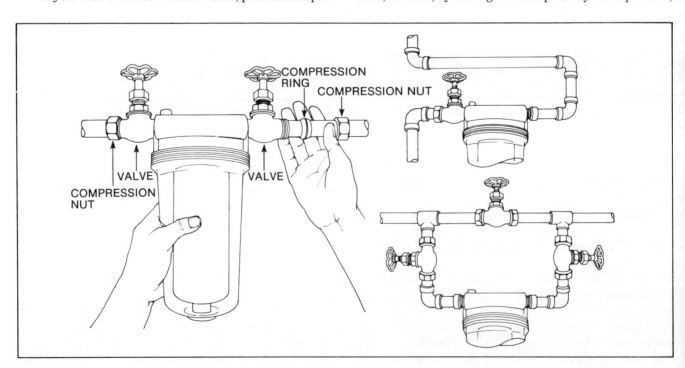

Installation of a water filter on a horizontal line.

one hot and one cold. Seal them in a jar for two or three days. If the sediment settles at the bottom, you probably need a filter to protect your pipes from buildup.

Other problems include hydrogen sulfide in the water, which produces a rotten-egg smell, or carbonic acid, which turns copper fixtures green. A small amount of these materials can be removed by softeners and filters. If there is a large amount, special equipment is necessary.

Water filters take care of sediment, algae, hydrogen sulfide, and other problems. To install a filter in a horizontal water-supply line, proceed as follows:

1. Shut off the water supply.
2. Cut out the section of pipe where the filter will be positioned. Allow room for a shutoff valve on both sides. By installing these valves you will be able to replace the filter every six months or so, without turning off the whole water-supply system.
3. Insert the valves into the filter cap and tighten them until they are in an upright position.
4. Connect the valves to the supply line with special plastic connectors or compression rings. When you are sure all connections are tight, turn on the water, then open both valves.

POSSIBLE CAUSE	REMEDY
Too few regenerations programmed.	Increase number of regenerations.
Salt in salt storage tank caked or bridged.	Remove salt bridge.
Increased water usage . . . guests, babies, etc.	Reprogram regeneration schedule.
Using hot water while softener is regenerating causing hard water to fill water heater.	Use hot water sparingly when softener is regenerating.
Possible increase in water hardness.	Take sample of water for test.
Leaky faucet or toilet valve. A leak the size of a pin can use over 100 gallons per day.	Repair to stop leak.
No salt in storage tank.	Refill with salt. Program a regeneration for same night.
Softener power cord unplugged from electrical outlet.	Reset day dial. Then plug in power cord.
Dead electrical outlet because of blown fuse or switched circuit.	Replace dead fuse or check for switch on circuit. Reset day dial.
No regeneration programmed.	Program regeneration.
Clogged nozzle and venturi.	Clean nozzle and venturi.
Loose brine line.	Tighten all line connections.
External bypass valve or valves improperly set.	Properly set bypass valve(s) for service.

The filter must be upright to do its purifying work. To install a filter in a vertical supply line, shut off the water to the line and cut out a 4- to 5-inch section. Construct a loop using elbows and short sections of pipe. Install the purifier in the lower half of the loop with the shutoff valve on the side where water enters the filter.

The core must be replaced every six months to one year, depending on the volume and quality of the water. To replace the core, shut off the valves, unscrew the filter body, remove and replace the core, then screw the filter body back onto the cap.

Larger water filters are installed much like a water softener. Install it before the softener to remove impurities that might clog the softener lines.

Testing the System

The water-supply system can be tested in the same way as the drainage system, but only potable (drinkable) water should be used. Also, it should be under pressure at least equal to the working pressure of the system, but not less than 60 pounds per square inch. A pump and a pressure gauge will be needed to make the test.

The following table gives several tips and a few fast checks which you can do that will often avoid an unnecessary service call. If your water softener fails to operate properly, check these things before you call for service.

Installation of Fixtures

Connecting the plumbing fixtures is often the final step in the installation of a home plumbing system. Fortunately, hooking-up toilets, sinks, bathtubs, and other types of fixtures to the water-supply and drainage lines of the plumbing network poses no major problems. In fact, this finishing work is often the easiest and most satisfying aspect of a major plumbing job. This also applies to smaller-scale jobs, such as the installation of new fixtures in place of old ones—the simplest way of revitalizing an aging bathroom, kitchen, or workroom.

Planning Connections

Kitchens, bathrooms, and workrooms in the home are all heavy traffic areas—used daily, and used hard. These functional centers of the home also happen to be the rooms where you will find the most plumbing fixtures—sinks, bathtubs, toilets, lavatories, and others. It quickly becomes obvious that the greatest care should be given to the selection, location, and installation of these fixtures. The wide variety of fixture styles available allows the home plumber great freedom to exercise his or her personal taste. Take the time to choose good quality fixtures which will complement both room size and style. This does not necessarily mean spending vast sums of money. Consider the purchase of new fixtures an investment, one which will pay off in years of trouble-free service. In particular, kitchens and bathrooms deserve special attention. Whenever possible, for example, arrange for back-to-back installation if your fixtures are to be in two rooms. For instance, the same lines can serve both the kitchen fixtures and the bathroom fixtures.

The Kitchen Sink and Its Appliances

While the kitchen sink may often be the only plumbing fixture in the kitchen, it is probably used more often, for more things, and in more ways than any other item in the room. Locating it in the best possible spot is vital to the whole kitchen plan. For instance, if there is a window at least 40 inches above the floor (enough to allow for a backsplash), you may want to place the sink under it. If there is a peninsula dividing the work area of the kitchen from the dining area, this makes an excellent site for a sink. Gourmet cooks often favor putting the sink in an island work counter in the middle of a kitchen.

Sinks come in three basic types:

1. Enameled cast-iron sinks are manufactured with a heavy wall thickness of iron; a fused-in enamel is applied to all exposed surfaces. The enamel on cast-iron sinks is four times thicker than on other types of sinks. This provides much greater resistance to cracking, chipping, and marring. Cast-iron sinks are available in colors or in white. The solid, heavy construction makes them less subject to vibration and, therefore, extremely quiet when installed with a disposal.

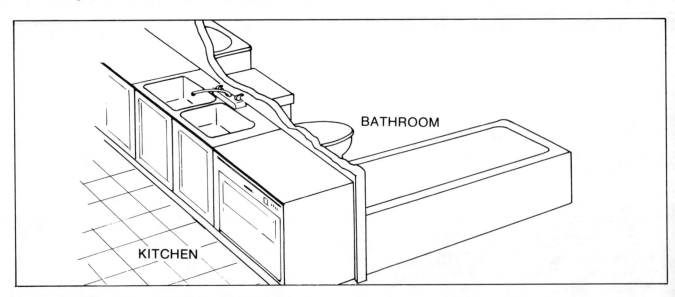

Back-to-back installation of fixtures will always save money.

2. Porcelain-on-steel sinks are formed of sheet steel, in one piece, and are sprayed and fired to produce a glass-like finish much like the surface found on cast-iron sinks. But, because of the unique physical characteristics of the material used, the finish on porcelain-on-steel sinks is only one-fourth as heavy as that on cast-iron sinks.

3. Stainless steel sinks are the lightest of the three and will provide a lifetime of service. The surfaces are easy to keep clean and are stain resistant. Since stainless steel is the finish, this type of sink is not available in colors. But the natural finish blends well with most color schemes.

In these three materials, there are single bowls, double bowls, deep bowls, shallow bowls, big bowls, little bowls. There are sinks with twin compartments set at an angle for use in corners; or you can simply place a conventional sink diagonally across a corner if that location is indicated.

The final selection of style and color is up to you. Most people prefer the double bowl, but it is wise to remember that one large sink (21 to 24 inches) takes less space than a double one (33 inches, for example) and is quite satisfactory if there is a dishwasher. Most sinks are 21 or 22 inches from front to back, and the bowl itself usually is 16 inches from front to back, but these dimensions vary slightly according to manufacturer and style.

While the usual depth of a standard bowl is about 7½ inches, this can vary to some degree, too. For instance, a triple-bowl sink may have a small vegetable sink between two standard bowls, and this may be only 3½ inches deep. In kitchens where space is at a premium, there are sinks at a depth of 5½ inches which permit one bowl to fit over a built-in dishwasher.

Frequently, if space permits, it is a good idea to have two separate sinks in the kitchen for two work centers, or an extra small sink for father's bartending, mother's plant watering, or the kids' fingerpainting. Also consider installing a sit-down sink. This is a sink that puts the homemaker close to the work and takes her off her feet at the same time. It makes no difference if she is short or tall; the sink can be installed at the height most convenient to her.

Sinks may come with faucets and other attachments or simply with punched holes so faucets can be purchased separately. One of the more popular sink attachments is an undersink heater, which serves at a tap instant near-boiling water for coffee, dehydrated soups, gelatin desserts, and the like.

Allow 24 inches on one side of the sink for a dishwasher. Be sure to avoid placing the dishwasher at immediate right angles to the sink; if it must be around the corner from the sink, place it 24 inches from the corner so you can stand at the sink and pivot to load the open dishwasher. You will need at least 24 inches of counter space on each side of the sink. There should be storage space nearby for hand towels, dish towels, dish cloths, and aprons; vegetables such as onions and potatoes; a chopping board; vegetable peeler; pots and pans used for foods cooked with water; and other items used first at the sink.

The Garbage Disposal A food waste disposal installed under one sink will eliminate messy food waste and simplify clean-up after meals, since you can peel fruits and vegetables, scrape plates, dump coffee grounds, and empty cereal bowls directly into the sink. The quality disposal can handle bones, fruit pits, egg shells, shrimp shells, in fact, almost any food waste. Some of the more efficient models have cutter blades that can handle fibrous waste such as corn husks, celery stalks, and artichoke leaves. Disposals are so efficient in handling food waste that today approximately 100 cities require their installation in all new housing starts or major improvements. Research by the United States Public Health Service has pointed out that septic tank-soil absorption systems that meet the Federal Housing Authority's minimum property standards can handle the additional loads from food waste disposals. The addition of ground food waste may reduce the time between tank cleanings by approximately one-third.

There are two basic types of disposals: the batch-feed and the continuous-feed. The former is controlled by a built-in switch. Waste is placed in the chamber, the cold water is turned on, and the lid is put in place. In some cases, the lid must be turned or positioned a certain way to start the motor; in others, a sealed-in magnet activates a hermetically sealed switch so that the disposal starts as soon as the lid is dropped into place. A continuous-feed disposal is controlled by a separate electrical switch installed nearby. It is a little less expensive than a batch model, but the cost of installing the switch almost equalizes the price. With a continuous-feed model you can feed waste into the disposal as it is operating.

Most garbage disposals are designed to fit any single or double sink with a 3½-inch to 4-inch opening and require a 120-volt AC outlet. You should check local plumbing codes and requirements to determine the amount of fall necessary and the method of connection. Detailed instructions for installation are provided with each model and, of course, they should be followed to the letter.

Automatic Dishwasher The installation of a built-in dishwasher is not difficult, and some manufacturers are supplying simple instructions for do-it-yourself installation where the local code permits.

A typical space requirement for a built-in dishwasher is 34 inches high, at least 24 inches wide, and 24 inches deep. The dishwasher fits into the space of a standard 24-inch base cabinet. In an existing setup, if there is a 24-inch base cabinet that can be removed, well and good. If not, it is possible to take out a wider cabinet or two cabinets that total at least 24 inches and use fillers on each side to fill the remaining space. Fillers are usually available from cabinet suppliers and some lumber yards. When existing cabinets are built on the job, some carpentry or cabinetmaking may be needed to open the space and finish the adjacent cabinets. Do not place the dishwasher at an immediate right angle to the sink. If the dishwasher must be around the corner from the sink, allow an intervening space of at least 18, and preferably 24, inches from the corner. Also, because the refrigerator generates cold and a dishwasher produces heat and steam, these two appliances will last longer and work better if they are separated by a 3-inch insulated filler strip.

A dishwasher takes a 120-volt, 60-hertz AC individual circuit, fused for 20 amperes. Three-wire electrical service to the dishwasher is recommended for connection to the terminal block and for grounding. To avoid the hazard of electrical shock, the dishwasher must be installed before it is used.

The water supply needed is 140 to 150 degrees F hot water at 15 to 120 pounds per square inch pressure. The water pipe should be ½-inch outside-diameter copper, with a ⅜-inch female pipe thread connection at the valve. A ½-inch by ⅜-inch male compression elbow is provided as an accessory. An 8-foot flexible drain hose with a ½-inch inside diameter is furnished. It is not recommended that the drain line be extended beyond the length of the hose provided, but should this be necessary, attach the hose to a line of larger inside diameter.

The most desirable drain system for a built-in dishwasher is through a drain air gap mounted at the sink or at countertop level. This accessory protects against a siphoning of the wash or rinse water and also prevents the possibility of food waste entering the dishwasher in the event of a plugged drain line.

Bathroom Planning

Since all bathrooms contain several fixtures—tub, toilet, lavatory, and often a shower stall or unit—precise planning is essential. Draw up a floor plan to exact scale. This will help avoid several problems, including lack of clearance space between fixtures or miscalculations concerning the amount of space necessary to move about the bathroom comfortably.

Another primary consideration is arranging fixtures for the least number of piping runs required. If the toilet and sink can be located on one wall, they may be able to be serviced by the same soil stack and vent. Some other points to remember when planning a new bathroom are:

- Shower stalls can be used in place of tubs to provide more floor area.

- Fixtures are available in many colors, so you can get away from the traditional white.
- Try to locate entrances to bathrooms in a position not seen from a living room or a dining room.
- Be sure to provide adequate ventilation and lighting.
- The bathroom door should be positioned so that, if suddenly opened, it will not strike a person using one of the fixtures.
- If the family is a large one, consider the advisability of partitioning the room with separate doorways so several can use the facilities in private.
- It is thrifty to arrange all of your fixtures in a straight line along a wall for this saves on pipe and fittings. Similarly, for a two-story house, locate fixtures in a continuous vertical line.

For a complete guide to bathroom planning, see *Bathroom Planning and Remodeling*.

Installation of Bathroom Fixtures

When installing bathroom fixtures during a construction or complete remodeling job, connect the tub and/or the shower stall before finishing the walls and floor. Afterwards, finish the walls and floor, then install the toilet and lavatory.

If you are replacing old fixtures with new ones, you may have to open the walls for installation of the tub and lavatory. If so, check the roughing-in dimensions of the new fixtures against the branch-drain and supply-line openings (already there) to determine how the fixture drains and supply lines will connect with them. Also, refer to the instructions furnished with each fixture. The drawing shows a typical branch-drain and water-supply openings (both in-wall and in-floor types) in relation to bathroom fixtures.

Bathroom fixtures are made of vitreous china, enameled iron, enameled steel, and fiberglass. Vitreous china is always used for toilets and may be used for lavatories. Porcelain enameled cast iron and pressed steel are used for tubs and lavatories. All white and colored china and porcelain enameled fixtures now on the market are acid-resistant.

In recent years, good-quality fiberglass fixtures have become available. The gloss and color of the gel coat finish are similar in appearance to those of enameled or china fixtures, and the finish is resistant to ordinary household chemicals. Tub-shower units and shower stalls, including the surrounding wall area, are of leakproof one-piece construction.

Bathtubs Standard bathtubs are available in a number of various designs. When considering styles of tubs, take into account some of the features manufacturers build into a product. For example, as a safety feature, some tubs have a textured, slip-resistant bottom surface. Another feature is a built-in electric, but

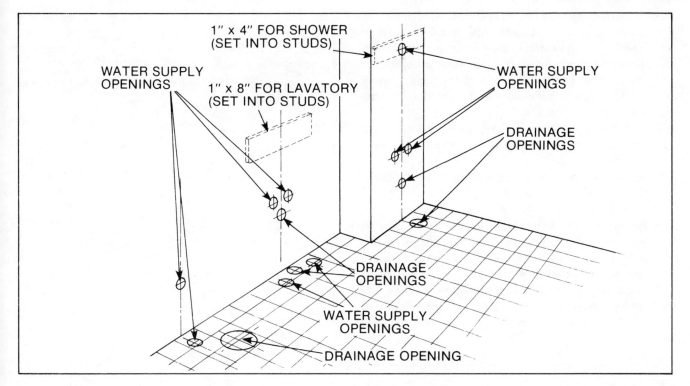

Basic information required to make proper connections of bathroom equipment.

safe, whirlpool system for therapeutic hydromassage. Some tubs have a raised outlet feature. These bathtubs, when used with wall outlet toilets, can reduce overall construction costs in certain situations, since all supplies and drains can be roughed into the wall.

The most economical way to provide a shower is to add a shower head over the tub. Separate stall showers are available with glass or plastic doors or are designed for shower curtains. Some types are so arranged that neither a door nor a curtain is necessary, since they are designed to keep the water and spray within the stall.

To insure head clearance for adults, the shower head should be installed at least 6 feet, 2 inches from the floor. Tubs installed with shower heads can be enclosed with permanent rigid enclosures or with shower curtains. Install the rod for the shower curtain at a height of 6 feet, 6 inches.

The Toilet Toilets (or water closets) fall essentially into five categories: (1) one-piece (neat in appearance and easily cleaned, but much more expensive that two-piece models); (2) close-coupled tank and bowl (the tank, a separate unit, is attached to the bowl); (3) two-piece with wall-hung tank; (4) wall-hung (completely wall-hung toilets make it possible to clean the floor under and around the toilet); and (5) corner toilet (a great space saver).

Any toilet you plan to install will operate with one of the following three flushing actions:

- Washdown. The washdown toilet bowl discharges into a trapway at the front, and it is most easily recognized by a characteristic bulge on the front

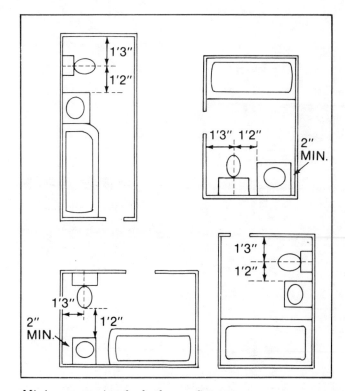

Minimum spacing for bathroom fixtures.

exterior. It has a much smaller exposed water surface inside the bowl, with a large flat exposed china surface at the front of the bowl interior. Since this area is not protected by water, it is subject to fouling, contamination, and staining. The trapway in the washdown bowl is not round, and its interior is frequently irregular in shape due to the exterior design and method of manufacturer. Characteristically, the washdown bowl does not flush as well or as quietly as other bowls. The washdown bowl is no longer accepted by many municipal code authorities, and several manufacturers have deleted it from their manufacturing schedule.

- Reverse trap. The reverse trap bowl discharges into a trapway at the rear of the bowl. Most manufacturers' models have a larger exposed water surface, thereby reducing fouling and staining of the bowl interior. The trapway is generally round, providing a more efficient flushing action.
- Siphon jet. The siphon jet is similar to the reverse trap in that the trapway also discharges to the rear of the bowl. All models must have a larger exposed water surface, leaving less interior china surface exposed to fouling or contamination. The trapway must be larger, and it is engineered to be as round as possible for the most efficient flushing action.

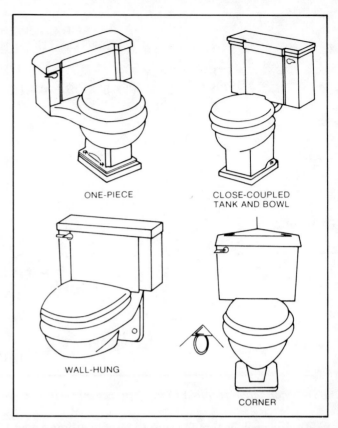

Common toilet (water closet) designs.

Lavatories Bathroom lavatories, like kitchen sinks, are available in a wide variety of styles and materials suitable for any type of bathroom layout. The many types of ornate fittings and decorated bowls allow you to give your bathroom precisely the atmosphere desired. Laundry tubs, workroom sinks, and the like need not be so elaborate. Let quality be your guide when selecting these fixtures.

After taking the time and effort to choose the right fixtures, be very careful not to scratch the finished surfaces when handling them. Rest the fixtures on pads whenever necessary, and use wooden blocks to brace them in position. Never stand on or in a fixture with your shoes on.

Installing a Toilet

There are two basic types of toilets: those with a one-piece tank and bowl design and those having a separate tank and bowl. In the installation of both types, the bowl is firmly bolted to the floor. This creates some unique problems. It is impossible to reach under the floor to tighten the final connection. Also, the connection may have to be broken for future repairs or the installation of a new unit. In either case, since punching a hole up through the ceiling beneath the toilet is most often impractical, the method of sealing the bowl to the drain line that connects to the drain main stack is unique.

For setting the bowl (any type), most plumbing in-

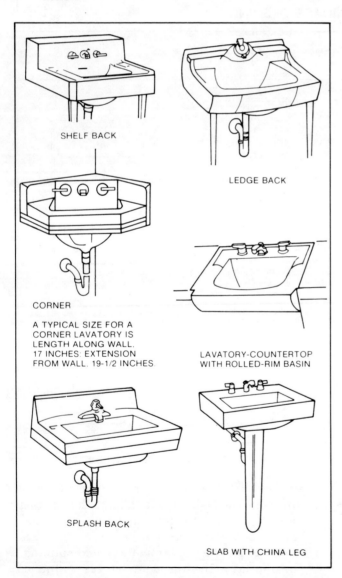

Common lavatory designs.

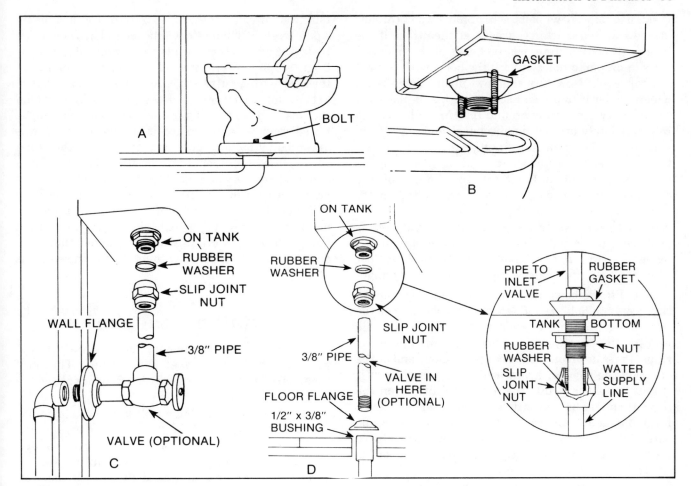

The installation of a toilet and tank.

stallations include a floor flange that rests upon the floor around the opening of the drain line. Be certain the toilet flange is located the correct distance away from the wall for the type you wish to install. Allow for both bowl and tank size. Twelve inches is usually right for most installations.

These floor flanges can accommodate two upright bolts. To set any bowl that is held only by these bolts, simply place the bolts in the flange slots provided. However, if your bowl requires four bolts, locate it on the floor properly over the flange, mark the spots for the two additional (front) bolts, then set these bolts into the floor at the positions marked. (If the floor is

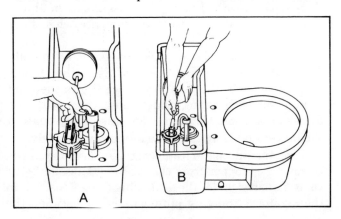

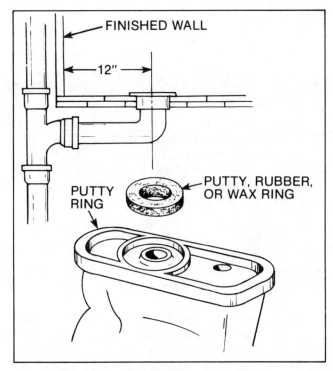

Preparing for toilet installation.

Final steps of installing a toilet.

wood, use toilet-bowl bolts, which have wood threads at one end and machine threads at the other. If the floor is tile or concrete, set the heads of the machine bolts in prepared holes and fill the holes with cement to floor level.) Now turn the bowl upside down (with papers under it to prevent scratching), and place one roll of putty completely around the rim and a second roll completely around the discharge opening. Putty rings should be about 1 inch high. Arrange the one on the rim to squeeze inward (by rolling it toward the inner edge of the rim), but arrange the one on the discharge opening to squeeze outward, so that putty cannot get into and clog the opening. Either a rubber or a wax gasket (both are made for this purpose) is preferred to the putty on the discharge opening. If necessary, two rings or gaskets can be used if the floor level has been raised above the top of the floor flange.

Now, lift the bowl up (it is heavy and may require two people to hold it) and set it gently down over the floor flange, putting it down as straight as possible into the final position, so as not to disturb the putty rings. Press down on the top center (not one end or the other) of the bowl with your full weight, and twist it slightly to settle it into the putty and firmly on the floor. It should be perfectly level when settled. Use a level, if in doubt, and wedge it up as necessary, but be sure that any wedging used does not lift it up to leave air gaps in either of the putty rings. When the bowl is squarely seated (with watertight seals at the putty rings), bolt it down to a snug fit, but do not forcibly tighten the nuts on the bolts. To prevent the accidental breakage of the fixture, alternate from bolt to bolt as you tighten.

With a separate tank toilet, the tank is set on the rear end of the bowl. It is usually held by two bolts, and the water connection is sealed with a fitted gasket. Fixture supply lines (which should be purchased to fit the installation) are connected.

To complete the installation, connect the plastic tubing which runs from the top of the water inlet regulator to a holding ring inside the overflow tube. Snap the plastic bolt tops into place on the bolts protruding through the tank base. Install the toilet seat and lid according to the manufacturer's instructions.

It is now possible to turn on the water supply. The tank should fill to the water level line marked on its wall. If it does not fill to the proper level, adjustments can be made by bending the brass rod which supports the float. Stop when the water reaches the marked water line.

When a toilet is replaced, the old unit, of course, must be removed before the new one is installed. To accomplish this, turn off the water supply at the main source and open any cold-water faucet to relieve pressure in the pipes. Flush the toilet to drain the tank. Remove the remaining water with a sponge. Disconnect the water-supply pipe from the underside of the tank; also disconnect the pipe at the wall or floor, using a pipe wrench. Then remove the old toilet. Temporarily stuff a rag into the sewer outlet to prevent sewer gas from escaping. Clean the old sealing compound from the floor with a putty knife. Remove the old wax gasket or compound from the closet flange or from around the waste outlet opening in the floor. If the flooring is rotted, remove and replace it. If the old bowl was anchored by bolts screwed into the floor rather than with a closet flange, check the position of the bolt holes and waste outlet opening of the new bowl to assure a proper fit with the existing bolts. If not suitable, they will have to be removed and new bolts installed, or plumbing work will have to be done to install a closet flange. Once this work has been completed, the procedure for installing the new toilet unit is the same as already described above.

Installing Lavatories

Lavatories are available in four separate types: wall-hung, pedestal, cabinet, and leg-stand. The top of a standard lavatory reaches 31 to 35 inches above the floor level.

If you plan on installing a wall-hung type lavatory, you will be using one of the three types of wall brackets. To support the (single or double) bracket solidly, provide a horizontal 1-by-8-inch board firmly anchored to two studs and embedded flush with the wall behind where the bracket will be. When the bracket is properly installed, the lavatory is simply hung on it (or them) against the wall. Make certain the bracket is secured firmly to the wall. It must be able to support the weight of the lavatory, the water in it, and any additional weight exerted on the fixture by a person leaning on its edge. If there are legs at the front, these will be adjustable so that they can be made to rest squarely on the floor. These legs also provide extra support. If a pedestal is used, this can be mounted on bolts like the toilet bowl. Cabinet types of lavatories do not require anchoring.

The fixture drain line for typical floor and wall connections must contain a trap to stop sewer gases from entering your home. Use of a shutoff valve in each (hot- and cold-water) supply line is optional, but highly recommended as a great convenience when later servicing faucets.

To install a basic lavatory faucet, assemble the rubber washer on each shank, then insert the shanks through holes in the lavatory. Place the washer and locknut on each shank and tighten them with the faucet in position. Connect the water-supply pipes to the faucet and tighten the coupling nut securely.

When installing the standard type of lavatory drain, remove the stopper from the drain and assemble the rubber washer under the drain plug flange. Insert it in the lavatory, applying plumber's putty between the

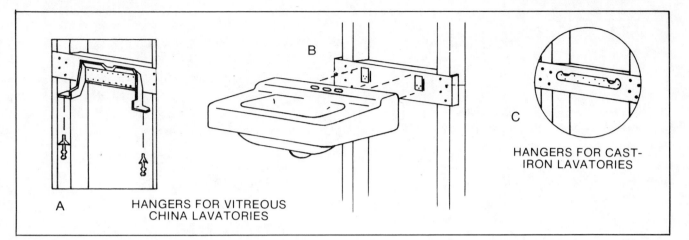

Three types of wall brackets.

lavatory and washer. Attach other washers and install the locknut. Thread on and tighten the pop-up assembly, using a good pipe joint compound. Make sure the lift rod points to the rear of the lavatory. Tighten the locknut to secure the drain in position. Insert the stopper. Then, loosely assemble the lift rod to the pop-up drain rod. Operate the pop-up drain and adjust it to position. Tighten the set screw to secure the linkage in the desired position. Tighten the sealing cap on the pop-up lever only enough to hold up the lift rod and knob.

To remove an old lavatory or vanity unit, turn off the water supply at the main source and open the lavatory or vanity faucets to relieve the pressure in the pipes. Drain the waste pipe trap (elbow) into a pail or pan by removing the plug from the bottom of the trap. Disconnect the waste pipe trap by unscrewing the large nut on each end of the trap with a parallel-jawed adjustable wrench. (If a drain plug is not provided, carefully disconnect the waste pipe trap and spill the water into a pail or pan.) Remove the waste pipe from the floor or wall nipple using the parallel-jawed adjustable wrench. Disconnect the water-supply pipes from the floor or wall nipples using the parallel-jawed adjustable wrench or an appropriately sized open-end wrench. The lavatory may now be lifted from its wall hanger. However, a vanity may require further disassembly. Inspect the underside of an enclosed lavatory or vanity top for attachments (screws or nuts) to the cabinet and remove them. Lift off the lavatory or vanity top. Inspect the back panel of the cabinet for attachments (screws or nuts) and remove these. Remove the cabinet. If a lavatory wall hanger is involved and found suitable for use with your new lavatory, both functionally and dimensionally, it may be left in place. If not, remove the attachments (screws or nuts) and hanger from the wall. Install a new hanger of the correct dimensions and proceed as described for installation of the new fixture.

Often, replacing a lavatory will remodernize the entire sink. To remove the old faucet, turn off the

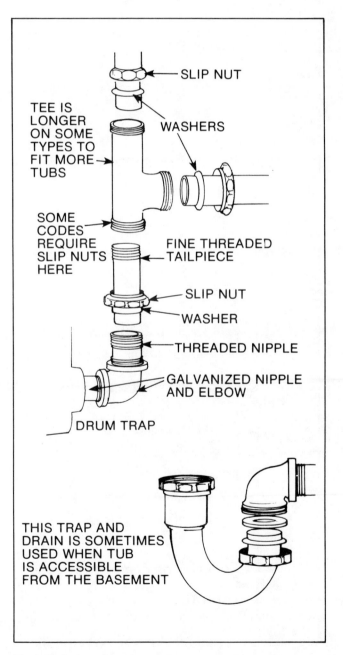

Typical drain connections.

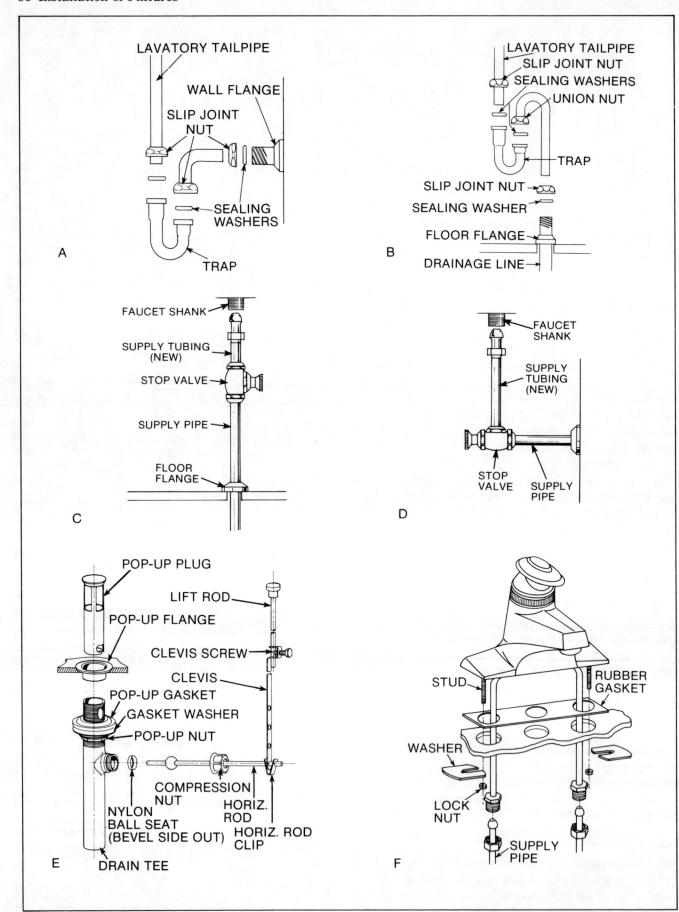

Necessary floor and wall connections.

water supply at the main source. Open any cold-water faucet to relieve pressure in the pipes. Disconnect the trap and swing the trap in either direction for additional working space. If the new faucet has a mechanical drain, remove the existing tail pipe. Disconnect the water-supply pipes from the old faucet with a basin wrench. Disconnect the supply pipes at the wall or floor using a pipe wrench. Using a basin wrench, remove the old faucet locknuts and lift off the old faucet. Use a putty knife to remove the old putty from the sink.

When installing a new lavatory faucet, install nipples and valves on the existing water lines. Turn the handle on the valves clockwise to close. At the main source, turn on the water so it is available to the rest of the house. Close the faucet that had been opened to relieve pressure. Refer to the instructions furnished with the new faucet before mounting it. Assemble the rubber washers on each shank; then, insert the shanks through the holes in the lavatory. Place the washer and locknut on each shank and tighten, using a basin wrench. Install the mechanical drain (if included).

To reconnect the supply lines, measure from the faucet shank to the shutoff valve for length of pipe required. If it is necessary to shorten pipes, the straight sections may be cut to within 1½ inches of the flexible portion. (This does not apply to floor supplies.) Using a hacksaw or tubing cutter, be sure to keep the cuts squarely at 90 degrees with the pipe wall. Remove sharp edges and burrs with a file. Slip the coupling nut (supplied with the faucet), compression nut, and ring on the supply pipe. Insert the straight section of the pipe squarely into the valve until it seats firmly. Tighten the compression nut over the ring finger-tight. Insert the nosepiece into the faucet shank. Be sure it enters straight and square. Slip the coupling nut into position and tighten. Force the straight pipe in the valve as far as it will go and tighten the compression nut with an adjustable wrench. Be sure to connect the hot water to the left side of the faucet and the cold water to the right side.

Installing Bathtubs

Modern steel and fiberglass bathtubs, designed to rest on the floor, completely fill the area (from wall to wall) in which they are located. For this reason, it is necessary to install the tub before finishing the walls and floor. If you are installing a new fixture in place of an old one, some disassembly work will be necessary.

When the space is just large enough for the tub, the wall covering (down to the studs) must be removed from the three adjoining walls. If the space is longer than the tub, with open walls at one end and the side only, then build the unopened end out to fill the extra space. After opening the walls, firmly nail 1-by-4-inch boards to the studs with their tops perfectly level and

exactly at the height required to rest the tub flanges on them. Lower the tub into position with the flanges resting on the boards, and anchor each end of the tub to the end boards with screws through the flange holes provided. The walls can now be refinished to make a neat fit around the tub. (Note: No boards are required for cast-iron tubs.)

Some building codes require that you have access to the tub's waste and supply lines. When the tub is located on the first floor, access can be gained through a 12-inch-long by 6-inch-wide opening cut through the floor at the head of the tub. Center the opening on the tub and extend it 3 inches into the area below the wall at the tub's front end. You can provide access to second floor tubs by installing a removable panel on the wall opposite the head of the tub.

The drain fitting of your tub will be similar to the one explained for lavatories. The tub branch drain (in the floor) should already contain a drum trap. All you need to do is connect the fixture drain to the branch-drain pipe in the floor below the tub. Connection is usually made with a slip-joint nut, the same as for the lavatory. Supply-line connections (also reached through the access panel) are also made in much the same manner as the lavatory connections.

The introduction of fiberglass units has greatly simplified the installation of tubs. In fact, the tub-shower unit is ready for installation once the framing and rough plumbing are complete. Standard framing techniques, using wood studs, are all that is required to house the 60-by-32-by-74-inch tub-shower. Depending upon working conditions, there are two alternate assembly methods of installing this unit. If the back or either end panel of the unit is inaccessible once the unit is in place, the components are assembled and fitted together and then set into the stud pocket. However, when the three wall sections can be easily reached, the base may be installed first. Wall sections are then set in place and sealed before the unit is fastened to the studs. In either case, caulking is easily done with sealant that comes with each unit. To complete the installation, the assembled unit is then lined up in place and fastened to wood studs with large-head galvanized nails or to steel studs with drywall screws. The final step is setting and installing the standard waste and overflow fittings in premolded openings, and the shower head, mixing valve, and spout.

To install the tub and overflow fittings screw each one into the special brass drainage fitting included with the tub fittings. Slip the 1½-inch tub drain pipe into the male threaded drain pipe which leads to the trap. Secure the connection with a 1½-inch slip-nut and rubber washer if you desire an accessible hook-up. Otherwise make a standard permanent connection.

The shower connection (if you are equipping the tub with one) is usually piped up from the faucet shower outlet with ½-inch pipe. A section of the chrome-plated pipe with the shower head extends out from a

90-degree elbow at the top of the shower pipe. Set a 1-by-4 board between the studs at the upper end of the shower pipe to help support it. Follow the manufacturer's specific instructions when adjusting the shower head, mixing valve, and spout.

Single-handle shower-tub controls are pressure-equalizing devices that not only mix the water, but also balance the water pressure so that, regardless of pressure changes in the supply lines, the water temperature at the spout or shower head remains constant. Because each make and model varies in design, it is impossible to give general directions on how to install them other than stating it is important to follow the manufacturer's instructions to the letter.

Installing Shower Stalls

The prefabricated shower stall units of molded plastic and of metal are very popular with the home remodeler because they can be used in place of tubs to provide more floor area. Shower stalls can be quickly connected to the water outlets and drains without danger of leaking from the sides and bottom of the fixture. They are available in sizes ranging from 32 by 32, 36 by 32, 42 by 32, etc., to 60 by 32 inches. Square units 36 by 36, 42 by 42, etc., can also be obtained. But, remember that only a 32-inch-wide, one-piece shower stall will pass through the rough framing of a 30-inch door.

Since the installation of each of these shower stalls varies with its manufacturer, no specific installation instructions are given here. The best advice that can be given is to follow the provided step-by-step directions to the letter. However, when a one-piece shower unit is installed, the pocket for the stall must be square and plumb, with the studs located as directed by the manufacturer. Since the one-piece unit combines the walls and base, no shower pan or hot mopping is required. The rough plumbing for the shower stall drain and water supply must be located in accordance with the dimensions given by the maker. Where access to valve connections is limited, consider using soft copper tubing from the risers to the valves from the supply lines. Connect the valves and risers before positioning the unit. The soft tubing will permit positioning of the valves and shower head after the unit is in place. For better floor support, apply a circle of cement or plaster about 18 inches in diameter on the floor around the drain pipe.

Once the shower stall is in place, level and plumb the unit. Then with number 6 large-headed galvanized nails, fasten the back wall of the unit to the studs, nailing through the nailing flange into the studs. Attach the side walls to the studs in the same way. If one end wall has been left loose, bring it up to the shower stall flanges and fasten it. The front vertical nailing flanges should be nailed to the studs on about 8-inch centers. Then the drain should be caulked and leaded. After the shower enclosure or shower curtain

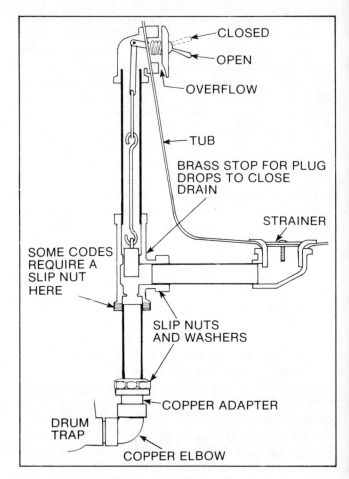

Typical tub fittings.

rod and other shower fittings are installed, the exterior of the shower stall pocket may be finished with the desired wall covering materials.

Installation of Other Fixtures

There are other fixtures in the home plumbing system besides those in the bathroom. But, remember that all fixture work is based on the basic procedures outlined on the preceding pages. By adhering to these points and any specific instructions given by the manufacturer, the home plumber can assemble fixture systems of professional quality and reliability.

Kitchen Sinks Kitchen sinks come in numerous varied designs of both the wall-hung and cabinet type. Cabinet designs are the most common. The method of mounting the sink depends entirely upon the kind of cabinet that holds it. Drain- and supply-line connections to a sink, however, are always the same in principle as those used for a lavatory. The removal procedure for an old sink fixture or faucet is also basically the same as the one outlined for lavatories. When you install a kitchen sink, a trap must be provided and, if you will drain the sink waste into a septic tank, a grease trap in the drain line is also advisable (sometimes it is required by the local code). If used, a grease trap may be inserted anywhere in the branch drain

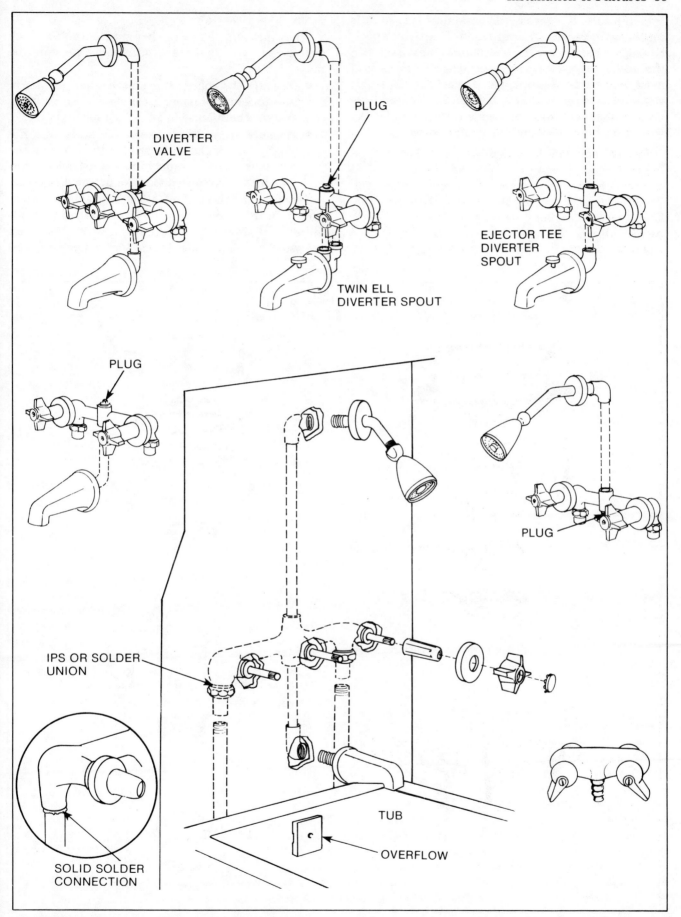

PLUG

DIVERTER
VALUE

PLUG

TWIN ELL
DIVERTER SPOUT

EJECTOR TEE
DIVERTER
SPOUT

PLUG

PLUG

IPS OR SOLDER
UNION

SOLID SOLDER
CONNECTION

TUB

OVERFLOW

Typical shower-head and tub fixture arrangements.

from the sink. If a garbage disposal unit is used, it is connected in the drain line above the trap (which must be exact in height, per the instructions contained with the unit), and the branch drain (clear to the stack) must be at least a 1½-inch pipe. If a garbage disposal is installed on one side of a two bowl sink, it should have its own separate trap and waste connection. Otherwise, water may back up the second drain.

To install an automatic dishwasher, follow the piping instructions supplied by the manufacturer. Most machines have air gap drains to prevent kitchen sink drain water from backing up into the dishwasher tub. Under some codes, these air gap drains are mandatory. As with all other fixtures, the dishwasher should have a trap of its own. Place it after the air gap. A branch drain is necessary also.

When installing a new kitchen sink faucet to replace an old one, two important measurements must be made:

- Measure the space between the stems (center of hole to center of hole). It can be 4, 6, or 8 inches, and your new faucet must fit your new or old sink.
- Measure the space between the stubs and the supply line. If the stubs are not the same on the new faucet, a new supply line can be installed or an adapter purchased to make up the difference.

The kitchen faucet is installed, as previously mentioned, in the same way as the dual-control lavatory faucet. The single lever faucet is also installed in an almost identical manner, except that holding bolts go

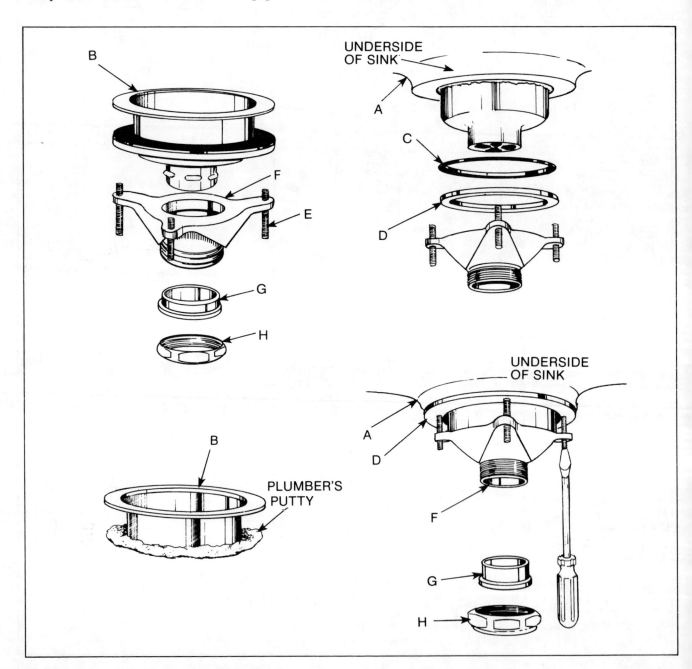

Steps in installing a typical kitchen sink strainer.

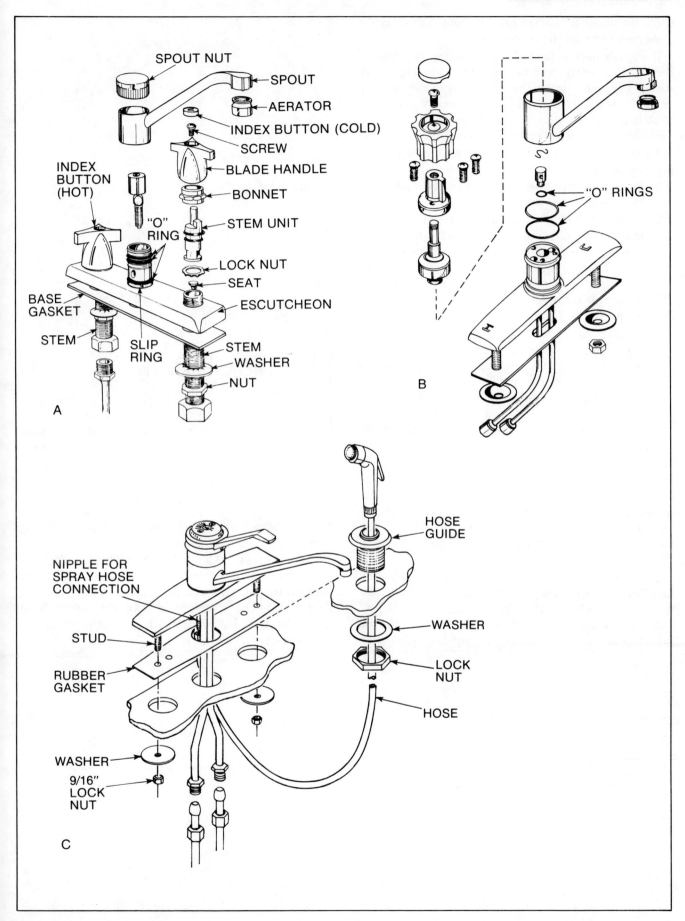

SPOUT NUT

SPOUT

AERATOR

INDEX BUTTON (COLD)

SCREW

BLADE HANDLE

BONNET

STEM UNIT

LOCK NUT

SEAT

ESCUTCHEON

INDEX BUTTON (HOT)

"O" RING

BASE GASKET

STEM

SLIP RING

STEM

WASHER

NUT

A

"O" RINGS

H

L

B

HOSE GUIDE

NIPPLE FOR SPRAY HOSE CONNECTION

STUD

RUBBER GASKET

WASHER

LOCK NUT

HOSE

WASHER

9/16" LOCK NUT

C

Typical kitchen sink faucets.

through outside holes. The hot- and cold-water-supply lines feed through the center hole.

If a spray hose is part of the fixture simply feed it through the fourth hole in the sink on the right and then to the center hole of the faucet. Apply pipe cement or Teflon tape to the threads at the end of the hose. Fasten the spray hose to the fixture and the spray assembly to the other end of the hose.

To replace or install a typical kitchen sink strainer, proceed as follows:

1. Loosen the three screws, twist the retainer, and remove from the assembly.
2. Insert the body through the opening in the sink, making certain that sufficient putty is used between the body and sink to seal properly.
3. Place the friction ring and washer in position on the retainer, and from the underside of the sink, attach the retainer to the body by turning it until engaged.
4. Tighten the three screws snugly, making certain that the body, retainer, and washers are lined up properly in the sink opening. Tighten the screws until the strainer is watertight.
5. Insert the sleeve into the retainer. Attach the tailpiece or tube to the retainer with a coupling nut. Remove any excess putty with a soft cloth.

Laundry Tubs Laundry tubs, washing machines, and other fixtures all have pipe connections similar to those already described in this chapter. When these fixtures are in the basement, it is sometimes desirable to use ordinary pipes and fittings (instead of chrome-plated ones) for the fixture connections. Always use traps of one kind or another when installing any of these fixtures.

Basement laundry tubs often differ from sinks and lavatories in that the water-supply lines may run down from overhead to the faucet connections. To provide air chambers, install tees directly above the faucet connections. Extend out from the sides with

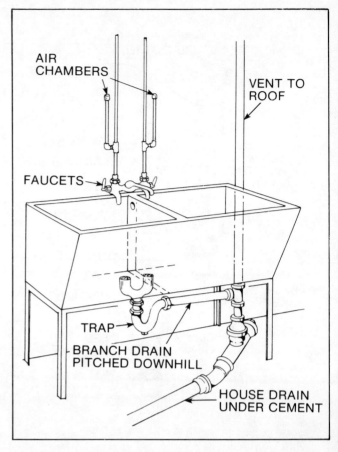

Installation of air chambers on a laundry tub connection.

short nipples, and then up with 90-degree elbows. Each of the two air chambers is made of a 12-inch-long capped nipple. These are fitted into the elbows.

Hose Faucets Located on the exterior walls of the house, brass hose faucets look and work in the same way as any other interior faucet. The water, however, flows inside the structure. For this reason, a frost-proof faucet, is ideal since it eliminates the need for an inside shutoff valve.

Installation of a Hot-Water Heater

A house plumbing system usually includes a water heater or a hot-water storage tank if the water is heated in the central heating plant. Electric, gas, oil-fueled, and solar water heaters are available. Each type comes in a variety of sizes. The tanks have the necessary internal piping already installed. The only connections required are the hot- and cold-water and fuel lines. Gas- or oil-fired water heaters require flues to vent the products of combustion.

Pressure and temperature relief valves are essential and should be on all water heaters and hot-water storage tanks. Their purpose is to relieve pressure in the tank and pipes if other control equipment fails and the water temperature goes high enough to generate dangerous pressure. As water heats, it expands; and the expansion may be enough to rupture the tank or pipes if the water cannot be forced back into the cold-water line or discharged through a relief outlet.

The "recovery rate" of water heaters varies with the type and capacity of the heating element. In standard conventional models, oil and gas heaters usually have higher recovery rates than electric heaters of similar size. However, a relatively new "quick recovery" type of electric water heater is available. Its two high-wattage heating elements provide hot water at a rapid rate.

The size of the hot-water storage tank needed in the house, depends upon the number of persons in the family, the volume of hot water that may be required during peak use periods (for example, during bathing or laundering periods), and the recovery rate of the heating unit. Household water heaters are generally available with tanks in a range of sizes from about 30 to 80 gallons. For a family of four or five persons, tank sizes should be about 30 to 40 gallons for oil or gas heaters, 40 gallons for quick-recovery electric heaters, and 40 to 52 gallons for standard electric heaters. For larger families, or where unusually heavy use will be made of hot water, correspondingly larger capacity heaters should be installed. Advice on the size needed may be obtained from equipment dealers and power company representatives. Incidentally, power suppliers may offer special reduced rates for electric water heating. If "off peak" electric heating will be used, be sure that the tank will hold enough hot water to last from one heating period to the next.

Location of Water Heater

Plan your supply lines to include the requirements for your hot-water heater. In most instances you will simply be replacing an older water heater in the same location. If you are considering relocation, the following should be kept in mind:

- You can locate an electric model where convenient, but a gas- or oil-burning heater must be placed within 8 feet of a chimney large enough for proper venting through the flue.
- Minimum clearances which must be maintained between the heater and combustible construction are: 1 inch at the sides and rear, 6 inches at the front, and 18 inches from the top of the jacket.
- National codes prohibit installation of gas water heaters in bathrooms or in any occupied room normally kept closed. Make sure there is adequate ventilation where such heaters are used. If a gas water heater is installed in an enclosed area (such as a utility closet), two ventilation openings for the replacement of air are necessary: one at the top and one at the bottom of the door or in the wall surrounding the utility closet. All openings must be a minimum free area of 100 square inches and have a width-to-height ratio of 2:1. The flow of combustion and ventilation air to the heater must not be obstructed. It is also necessary to have a way of replacing room air when exhaust fans are used.

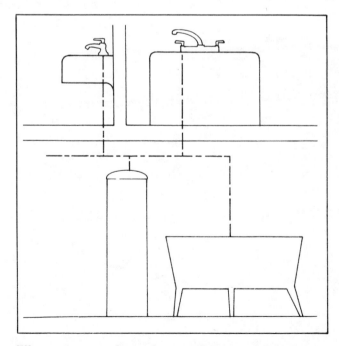

Whenever you can, locate the water heater as close as possible to the hot-water faucets it will serve. There will be far less heat loss and hot water will "come on" quicker.

- Put the heater as close as possible to where you use the most hot water.
- It is handy to have a floor drain, tub, or sink nearby. That will make it easy to drain water from your heater. It is also a good place to end the drain line of the temperature and pressure valve.
- The tank or the pipes and your connections may leak in time. Put the heater in a place where a water leak will not damage anything. The manufacturer is not responsible for any water damage.

- You must not put your heater in an area where it might freeze. You must turn off the electricity to the heater before you drain it, to protect the heating elements.
- Make sure that you are able to reach the drain valve and all access panels when your heater is in place. This will make it easy to service your heater.
- The heater must be level before you begin the piping.

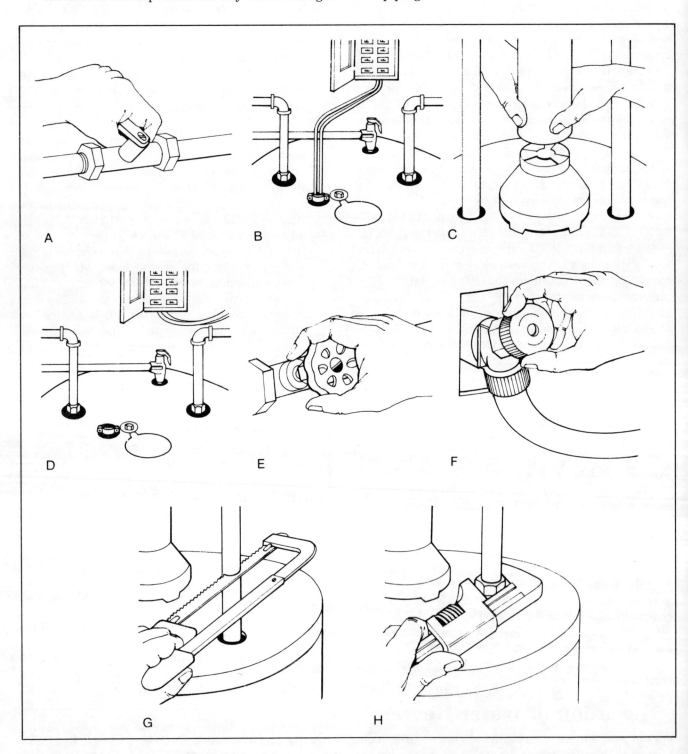

Steps in removing an old water heater.

In addition to water pipes, you must install gas piping for such heaters. If you are installing a water heater of the same size as the one replaced, all that is involved is reconnecting the gas line. If your replacement unit is larger, consult the installation guide packed with the unit for the proper pipe size. Indoors, this pipe can be black steel or flare, or compression-type copper (never soldered copper). However, copper should not be used inside of walls or in any inaccessible areas. Outdoors, gas pipe can be made of threaded steel (coated and wrapped), certain grades of copper, or plastic. Check your local codes for the proper material to use.

Removing the Old Heater It is necessary to completely remove the worn-out heater. This will enable you to install the new heater in a larger working area with all fittings more accessible. To remove the heater:

1. Turn off the gas supply at the meter (A) or turn off the electricity at the fuse box (B).
2. Remove the vent cap (C) if it is a gas heater, or disconnect the heater's electric source (D).
3. Close the valve on the cold-water supply (E). Open a hot-water faucet to allow air to enter the system.
4. Connect a standard hose to the drain cock; empty the tank (F).
5. Carefully disconnect all fittings. If the supply lines are copper tubing, cut them 4 inches above the heater (G). If they are galvanized steel, unscrew pipe at the unions (H). Remove the old heater.

Purchase the necessary fittings to adapt existing plumbing to the new heater and to connect the cold- and hot-water supply pipes to the heater (identified by a marking on the top). Be sure the cold-water inlet connection contains the factory-furnished dip tube. The cold-water inlet line should contain a shutoff valve and union. The hot-water supply line should also contain a union.

Installing a New Water Heater

Gas and electric water heaters are both connected into your supply system in the same manner. The actual choice of heat source should be based upon which is the most economical in the area where you live.

Gas Hot-Water Heater Instructions for connecting a gas water heater to the plumbing system come with the unit. But, in general, the following steps are involved in installing a gas hot-water heater:

After you have moved the new gas water heater into the desired location and made sure it is level, take a close look at the top of the unit to identify the hot- and cold-water connections. The factory usually supplies a tank with a dip tube to carry the incoming cold water to the bottom of the tank; hot water rises, so naturally

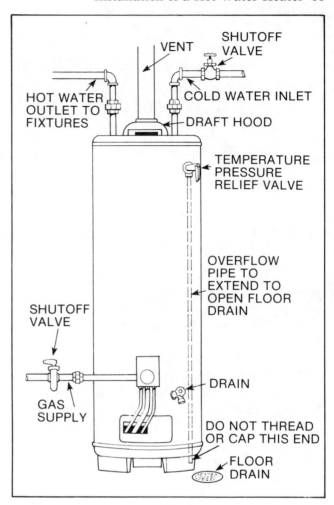

Typical gas hot-water heater and its parts.

it exits from the top of the tank. If the right and left positions of hot and cold connections seem to be opposite what would be the most convenient for your application, you can switch them by moving the dip tube to the opposite side.

Before actually connecting the cold-water supply, install a shutoff valve in the line. This will facilitate periodic tank draining and make it possible for your family to at least have cold running water should your heater require servicing at a later date.

Measure the distance between the nipple fittings on top of the new heater and the existing water lines above. If the water lines are copper tubing:

1. Cut the tubing to the required length for both the hot- and cold-water connections. Use a tubing cutter.
2. Solder one end of each tubing length into a ¾-inch adapter and connect it to the heater nipples. (Be sure to do soldering away from the heater as some nipples contain a plastic liner.)
3. Insert the opposite end of each tubing length into the adapter on the existing piping and solder the joint. It is usually desirable to replace the existing shutoff valve in the cold-water line. Do not

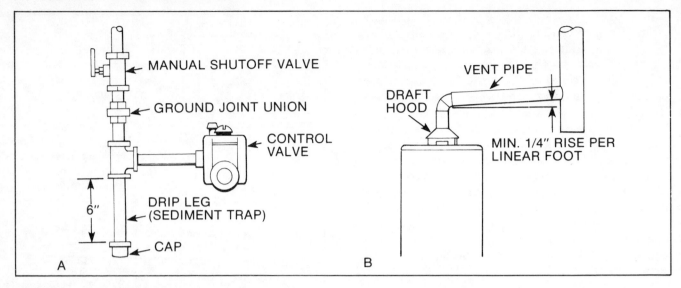

MANUAL SHUTOFF VALVE

GROUND JOINT UNION

CONTROL VALVE

DRIP LEG
(SEDIMENT TRAP)

6"

CAP

A

VENT PIPE

DRAFT HOOD

MIN. 1/4" RISE PER LINEAR FOOT

B

(A) Gas piping and (B) vent pipe installation of a typical gas water heater. The vent pipe from the heater to the chimney (vent stack) must be no less than the diameter of the draft hood outlet on the heater and should slope upward to the chimney at least ¼-inch per linear foot.

put a shutoff valve in the hot-water line.

If the water lines are threaded pipe:

1. Use threaded ¾-inch pipe to the required lengths for both the hot- and the cold-water connections.
2. Join the pipes to the nipples on the top of the heater using unions.
3. Join the opposite ends of the pipes to the existing piping with unions.

If you are connecting your new heater to a CPVC or PB plastic system, make certain the plastic hot-water line is connected to a galvanized pipe at least 18 inches long from the outlet side. While plastic is rated at above 100 psi at 180 degrees, the section of galvanized pipe protects the plastic from higher temperatures that may be transferred from the tank itself.

You must be sure that all gas piping meets the local code or gas utility requirements. Measure the vertical distance from the gas shutoff valve (gas cock) to a point opposite the gas control. Connect ½-inch length of black iron pipe to the gas shutoff valve. Then, measure the horizontal distance from the gas control to the vertical pipe. Use a union and two lengths of black iron pipe to fit. Attach the drip leg and cap to the tee at the junction of the vertical and horizontal pipes.

To install the vent hood and to make the connection to the flue, carefully check the manufacturer's instructions. Once the draft hood is installed, position the water heater so that the vent pipe will fit over it. With the vent pipe in position, drill a small hole through both the vent pipe and draft hood. Secure them together with a sheet metal screw.

Apply Teflon tape to the male threaded end of the new temperature and pressure (T&P) relief valve and screw it into the opening marked "relief valve." (Never reuse the T&P valve from the old heater.) Tighten it with a wrench. Then, run a pipe from the relief valve

outlet (the pipe must be the same size as the outlet) to a suitable drainage point. Leave about a 6-inch air gap between the end of the pipe and drain. Do not install a shutoff valve in the relief drain line. Also, do not thread, plug, or cap the end of the relief pipe.

After checking to be sure all connections are tight and the drain valve near the bottom of the water heater is closed, you can turn on the incoming water to test your tank and water system for leaks. The air which is in the tank is being replaced by the water and must be allowed to escape. Open a hot-water faucet to allow this air to escape and continue filling the tank until the water runs continuously from the faucet. Turn the faucet off.

To test all gas connections for leaks, coat them with a soapy solution. If bubbles appear, a gas leak is indicated, and it will be necessary to tighten the connection. Never use an open flame for this test.

As a rule, directions for lighting the pilot are given on a metal plate which is attached to the side of the heater near the thermostat. Follow the procedure outlined below:

Turn the gas cock to the "Pilot" setting and hold down the reset button. By doing this, the reset button forces the interrupter to close the burner supply line. The same action pushes down the solenoid valve which opens the pilot supply line. The pilot can now be lit without the danger of the entire burner igniting.

A primary burner aid adjustment is required on all gas water heaters. This is accomplished by an "air shutter." The air shutter is located at the end of the burner venturi. The adjustment of air is made after you have lit the pilot light. Just remember: Open the shutter to sharpen the flame, close the shutter to soften the flame. The correct air adjustment will produce a soft blue flame.

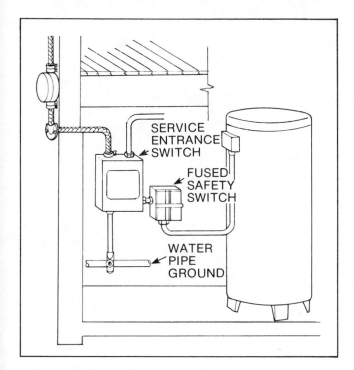

Typical electric water heater installation.

Make periodic inspections of the main gas burner, the pilot burner, and the venting system. The gas burner should have an even, all-around flame. In the venting system you should look for:

1. Obstructions which could cause improper venting.
2. Damage or deterioration which could cause improper venting or leakage of combustion products.

Whatever model gas water heater you use, be sure to follow these important safety precautions if you smell leaking gas:

1. Open the windows.
2. Do not touch electrical switches.
3. Extinguish any open flame.
4. Immediately call your gas supplier.

Electric Hot-Water Heater A double-element electric water heater is probably better for larger families because it permits a more constant supply of hot water. Double-element heaters have two thermostats. The single-element type has only one thermostat. The size of elements, type of thermostats, and method of wiring for heaters are usually specified by your power company. When the service panel has an unfused tap for a heater, use an indoor safety switch.

When making an electric water heater installation, follow the illustrated installation setup, with a fused safety switch and careful grounding. As mentioned earlier, power companies usually offer a very low rate for current used to heat water. This is known as an "off-peak-load" rate and is offered during hours when the demand for current is not great. The company will install a separate meter and time switch, which turns

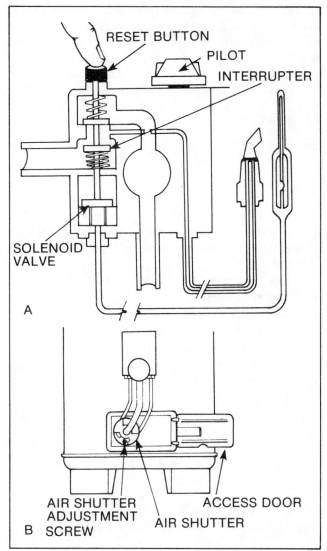

(A) Method of lighting the pilot; (B) method of adjusting the burner flame.

on or off at certain hours.

Position the water heater so that the existing piping and electrical hookup will require the shortest distance between connections. If you are replacing a water heater of the same size, reconnect the electrical line at the access on top of the heater. If your new heater is larger than the old one, a new properly-sized individually fused circuit may be required at the main junction box. Observe the requirements of the National Electrical Code as follows: If heater wattage is 3500, use No. 12 wire and 20 amp fuse; if heater wattage is 5500, use No. 10 wire and 35 amp fuse.

All water connections are made in the same manner as for gas water heaters. After checking that all connections are tight and the drain valve near the bottom of the water heater is closed, the cold-water supply valve to the heater can be turned on. (This valve must be left open when the heater is in use.) The air which is in the tank is being replaced by the water and must be allowed to escape. If any air is left in the top of the heater, the top heating element will burn out right away. Open a hot water faucet to allow this air to escape and continue filling the tank until water runs continuously from the faucet. Turn the faucet off. Checking now that the electrical connections are proper and safe, the electricity can be turned on. The water heater is now operational and will supply hot water after the first hour.

Maintenance

Draining the tank once a month prevents excessive mineral buildup and lengthens the life of your heater. A water heater should also be drained if it is being shut down during freezing temperatures. Also, periodic draining and cleaning of sediment from the tank may be necessary.

1. Turn the gas cock to the off position or turn off the electrical supply.
2. Close the cold-water inlet valve to the heater.
3. Open a nearby hot-water faucet.
4. Open the heater drain valve.
5. If the heater is going to be shut down and drained for an extended period, the drain valve should be left open.

When the tank is fully drained, open and close the inlet shutoff a few times to flush the last of the sediment away, then close the tank drain and refill.

Insulating Your Water Heater Except for an uninsulated attic, an uninsulated water heater wastes the most fuel in the house. Fiberglass insulation can be used to cut down on fuel costs. You will need a piece that will surround the body of your heater, plus vinyl tape to seal the seams. If you do not want to measure and cut the insulation yourself, kits are available.

For electric heaters, the first step is to measure the diameter of the top and cut a top plate to size. Next, for both gas and electric, measure the length and height of

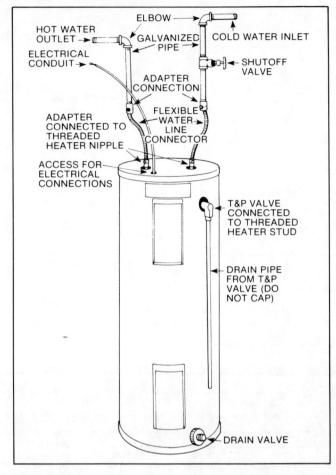

Typical single-element electric water heater with galvanized pipe, flexible connectors, and elbows.

the water heater and cut the insulation to fit the body. Wrap the heater with the insulation and apply vinyl tape to the seam. Caution: Do not block the air flow in the stack of a gas heater. This may increase the temperature and create the potential for an explosion or fire.

Private Water Supply and Disposal

Private water-supply and disposal systems are necessary in rural areas that city water and sewer lines do not service. A well is the most common water source, and a septic tank is the usual sewage disposal system. Also, some people dislike the taste of treated municipal water and, even though water is available, they drill a well for a private supply.

There are several advantages to private systems. You do not have to pay monthly bills for either water or sewer; however, the water system requires electricity. Both wells and septic tanks, if installed correctly and according to code, should last a long time without repair. A few routine maintenance jobs will be necessary every several years, such as disinfecting the well or pumping out the septic tank.

Most of the work described in this chapter is best left to a professional. For example, it is up to a well contractor to decide, depending on the characteristics of the soil in your area, whether you have to drill as little as 30 or as much as 600 feet to reach water. He will drill the well for you, test the gallons per minute it can supply, and recommend a type of pump. For a septic system, some home owners cut costs by having professionals dig the trenches and then install the rest of the system themselves. Where conditions are right, you can dig a shallow well or install a septic system on your own, but be sure to get the proper permits, if necessary.

Developing A Water Supply

Finding water can be a problem. Any surface water can be considered contaminated. Even a fast-moving brook in sparsely settled areas is liable to be polluted. Frequently, such a stream will test well on one day and show contamination on another. In fact, any underground water within 10 feet of the earth's surface can likewise be regarded as unsafe. Sometimes water found at depths of more than 10 feet is impure because of underground contamination. The only way to find out whether the water is safe to use is to have the water tested for purity.

The depth you will have to go is a questionable matter. It often happens that whereas your next door neighbor finds an abundant supply of water with a 28-foot well, you may have to go 30, 35, 50, or even 100 feet—and, sometimes, in spite of this, you end up with nothing more than a dry hole in the ground. (It is estimated that about 10 to 12 percent of wells dug or drilled are unsuccessful.) Usually, the local forest officer, the state soil-conservation department, or

county extension office, through the help of their water-table charts, can tell you the depth you will have to go for water.

The location of your well is of the utmost importance. It should be placed on upgrade and at least 100 feet from any sources of pollution, such as outside privies, garbage pits, surface streams, and drainage ditches. Sanitary conditions around a well already in service sometimes can be improved by relocating the sources of pollution.

Generally, wells are either driven, dug, bored, or drilled. Basically, each type is a hole deep enough to tap an underground water-bearing formation. Also, the protection for all types of wells is the same: a pipe or well lining deep enough to exclude underground seepage of contaminated water and a water-tight platform or cover to prevent surface water from running into the well. Check with neighbors, plumbers, well-diggers, and the local health department to find out which type of the three wells seems to offer the best possibilities for you.

Driven Wells This type of well is used when it is possible to get down to the water-bearing stratum without running into hard rocks. Driven wells are seldom over 100 feet deep. The well itself consists of a 2- or 3-inch extra-strength wrought-iron pipe with one end fitted with a drive-well point. The other end of the pipe is fitted with a drive head. The pipe is then sunk into the ground by striking the drive head with a sledgehammer. When one section of pipe has been driven in, the drive head is removed, another section of pipe is attached, the drive head is placed on the new section, and the work continues until the drive-well point strikes water. This method of making a well is not possible in rocky soil because the drive point cannot be driven through rock.

A concrete platform at the top of the well gives protection against seepage and provides a substantial base for the pumping equipment. The pipe should extend at least 1 inch above the concrete pump base. Bolts to provide anchorage for the pump stock and pumping equipment should be set in the concrete before it hardens. Allow the concrete to cure fully before installing any equipment on the platform.

Dug Wells In soils where water is not far below the surface, a dug or shallow well may prove adequate where the demand for water is not too great. The well should be dug only deep enough to ensure an adequate flow of water. The inside of the well is lined with rocks or tongue-and-groove concrete tile pipe, and it should be provided with a tight-fitting lid to keep out rodents

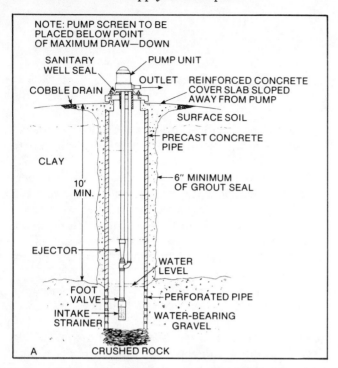

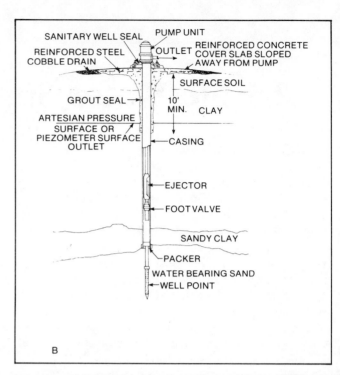

Three popular types of wells: (A) dug well; (B) bored well; and (C) drilled well.

and insects. A shallow well pump and a storage tank are usually required with a dug well system.

If you plan to dig your own well, the help of another person is required. One does the digging while the other hauls up the dirt and rocks in a bucket. Great care should be taken in doing this work because in certain soil conditions there is the constant and very real danger of cave-ins. Care also must be taken to avoid having any water in the well above the frost line, as resulting ice in the winter season will cause the cylinder to crack. In oiling the well machinery, allow no oil to seep into the well, or pollution will result.

Many old large-diameter wells are dangerous because polluted water seeps or drains into them. Many are lined with loose brick or stone, which offers little protection against seepage. These old wells can be protected and made serviceable by a concrete casing and watertight concrete platform. Actually, the same method used in protecting old wells is used for new dry wells. The old masonry casing is removed and a concrete pipe set in the well. The space between it and the earth walls of the well is filled with rock or gravel up to where the top casing begins. The platform is then constructed in the same manner as for new wells. After the well is protected, it should be pumped for a sufficient time to remove the contaminated water. Then, a sample of the water should be tested to make certain it is pollution free.

Bored Wells Bored wells are commonly constructed with earth augers turned by power equipment. Such wells usually are regarded as practical at depths of less than 100 feet when the water requirement is low and the material overlying the water-bearing formation has noncaving properties and contains a few large boulders. In suitable material, holes from 2 to 30 inches in diameter can be bored to about 100 feet without caving. In general, bored wells have the same characteristics as dug wells, but they may be extended deeper into the water-bearing formation.

Bored wells may be cased with vitrified tile, concrete pipe, standard wrought iron, steel casing, or another suitable material capable of sustaining imposed loads. The well may be completed by installing well screens or perforated casing in the water-bearing sand and gravel. Proper protection from surface drainage should be provided by sealing the casing with cement grout to a depth of at least 10 feet below the ground surface.

Drilled Wells By far, the most efficient private water supply in most cases is that obtained from a drilled artesian well. These wells are sunk deep into the ground to the point where they reach water-bearing stratum. This stratum may lie many hundreds of feet under the surface. The wells are expensive and cannot be dug without special power-drilling equipment. But, in spite of their high costs, drilled wells are usually the best, because they can be counted on to supply plenty of fresh water. They will not go dry during drought periods as will wells that are not very deep.

Regardless of your method of construction of the well, select the well-contractor most carefully. Ask for recommendations from the people in the area, the local health department, the state soil-conservation service, and the county extension service office. Watch out for so-called "sweetheart" arrangements between

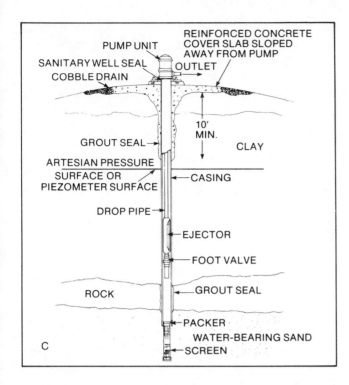

C

PUMP UNIT
SANITARY WELL SEAL
COBBLE DRAIN
REINFORCED CONCRETE
COVER SLAB SLOPED
AWAY FROM PUMP
OUTLET
10' MIN.
CLAY
GROUT SEAL
ARTESIAN PRESSURE
SURFACE OR
PIEZOMETER SURFACE
CASING
DROP PIPE
EJECTOR
FOOT VALVE
ROCK
GROUT SEAL
PACKER
WATER-BEARING SAND
SCREEN

a realtor or developer and a specific well-digger. The National Water Well Association (the association for the well-digging industry) recommends the following things to look for to help you judge the qualifications of a responsible well-digging contractor:

1. Willingness to give a list of previous customers. This will permit you to determine his reliability and reputation. Check also how long he has been established in the area and how modern his equipment is.
2. Willingness to work under a written contract with adequate liability insurance to protect you against a lawsuit in case of accident or damage.
3. Willingness to detail the job step-by-step and to itemize his estimate rather than provide a simple "per foot" figure. You should insist on both a certified log of the well and a statement of work performed. The log should cover the depth and thickness of the various layers of drift, rock, shale, and the like encountered during digging. A qualified contractor will gladly give this certification.
4. Willingness to guarantee the materials and quality of workmanship to your satisfaction. The contractor should report the results of pumping tests, the size of well-casing and screen, and the indicated water yield per hour.

The latter is most important. As a rule, a minimum of 50 gallons of water per day for each member of the family is the standard estimate. However, where there are dishwashers, clothes washers, and other water-using appliances, this demand figure should be increased upward to as high as 100 gallons of water a day

per person. Another point to keep in mind is that fire-insurance underwriters often demand a minimum yield from private wells for fire fighting. As a rule, their requirements are a stream of 10 gallons per minute through a ¼-inch nozzle, able to continue for two hours. (The total volume would be 600 gallons.) Remember that insurance rates can be very high if the well does not give adequate fire protection.

Water Pumps

A spring-fed water system can work on a gravity arrangement or can be operated in conjunction with a water pump. But, if you have a well, you need a pump to obtain "city-type" water service. You must have a pump that will deliver at all times the total maximum gallons per hour (gph) you will need. Water needs per day are based upon known average requirements (see the table), which are then converted by formula to arrive at the necessary pump capacity in gph.

WATER-REQUIREMENT CALCULATIONS
Number of Fixtures

1	Bathtub and/or shower
2	Lavatories
1	Toilet
2	Sinks and/or tubs
1	Automatic clothes washer
0	Automatic dishwasher
2	Garden-hose outlets
9	TOTAL

NOTE

Bold figures give an example of how to make the necessary calculations.

FORMULA: 9 TOTAL × 60 = 540 GPH PUMP CAPACITY.

Minimum pump capacity recommended for a home is 540 GPH.

For all practical purposes, a minimum discharge pressure of 30 pounds is sufficient to force water up to—and in good flow from—the highest faucet in a two-story structure. (Many city water mains deliver water to basement mains at 30 pounds or a little more.) If, however, the pump will be situated many feet below the basement of your home (pump and well in a valley, house on hill), a greater pressure will be required. The same applies if the pump will be at some distance from the building, or if it must deliver water to a distant outlet. In such cases, a larger pump and/or motor will be required—or you may need a booster pump in your system.

Every automatic well pump requires a water-storage tank, which is kept filled by intermittent operation of the pump. As the tank fills, the water compresses the air locked in the tank—and this air pressure will, in turn, force water from the tank up to and out an opened faucet. The pump must be equipped with an automatic pressure switch, which starts the pump when tank (air) pressure drops below the established minimum and stops the pump when it reaches an established maximum. (The pump must, of course, be able to deliver this maximum pressure at the tank.) A well-designed, glass-lined tank is essential if you want pure water and no external condensate to dampen the tank area. An ample-size tank also will guarantee plenty of instantaneous water, even in excess of pump maximum-gallons-per-minute capacity.

Nearly all manufacturers of pumps make several different models and types of pumps. Each may be classified as either a shallow-well or a deep-well pump.

A shallow-well pump is one in which the pumping mechanism, whether it be a cylinder, impeller, or jet, is located above ground but within a limited vertical distance of the surface of the water. This mechanism may be built in the pump frame or in the housing surrounding the pump and motor, or it may even be a separate unit.

A deep-well pump is one in which the pumping mechanism is usually installed below the surface of the water. For this type of pump, the cylinder, jet, impeller, or turbine is usually a separate and distinct unit from the motor or pump frame. There is also the submersible type.

Shallow-Well Pumps A shallow-well pump can be used in a spring or in a well where the water does not have to be lifted vertically more than 22 feet to get it to the possible suction lift, which is about 1 foot less. Often shallow-well pumps are located a considerable distance from the well or spring. Where the suction head will exceed 22 feet, a deep-well pump must be used. A jet-type deep-well pump may be located similarly to shallow-well pumps but must be connected to the well or spring by two pipes.

It is preferable that the pump be located in a frostproof, well-lighted, and ventilated place, such as the basement of the house, where it will be readily accessible for servicing. Frequently, pits or pump houses must be used for shallow-well pumps. Both should be large enough to permit easy access to the pump for servicing, and suitable drains must be provided to prevent water and oil from entering the well. Suction pipes for all offset installations must slope continuously uphill from the source of water to the pump. This is to prevent air from accumulating in the pipe. Under no circumstances should the size of the suction pipe be less than the size of the tapped hole in the pump housing. Smaller pipe will increase friction, lower efficiency, and may even prevent the pump from working at all.

The oldest type of shallow-well pump, and yet one that is still popular, is the reciprocating or piston-type pump. This type of pump is usually less expensive in the initial cost than most others. It will produce a relatively steady flow regardless of the pressure. Since it is a piston pump, it does have moving parts which will wear out and which will have to be replaced. This pump may sometimes be noisy and cause considerable vibration.

Deep-Well Pumps A deep-well pump is one that is designed to lift water vertically from depths exceeding 22 feet. It may be either the turbine type or the reciprocating piston type.

In a turbine-type pump the height of lift is governed by the number of stages and the horsepower supplied to operate the pump. It is self-priming, smooth-operating, and very flexible in both pressure and capacity. From a given size of well, it will deliver more water than any other type of pump. The turbine pump has a relatively higher initial cost.

The reciprocating plunger or piston-type pump is the simplest of the deep-well pumps. This pump is a positive-action type and is designed for use in wells of any depth up to 800 feet. It is quite efficient over a wide capacity range and its discharge is relatively constant regardless of the pressure. Either of these pumps must be placed directly over the well. That is, if you use a deep-well reciprocating pump or turbine pump, it will be necessary either to place the pump in a pit or build a frostproof house directly over the well. In either instance, a weatherproof door or small opening should be provided in the room directly over the well casing. Such an opening will permit the removal of the pipe when repairs become necessary.

Jet Pumps Ejector or jet pumps may be classified as either shallow-well or deep-well pumps. The main difference is in the jet assembly. For a shallow well, the jet assembly may be located near the pump or in the pump housing and obtains water by suction as does any other shallow-well pump. In a deep well, the jet assembly should be installed below the surface of the water.

The jet pump has increased in popularity in recent years to the point where it is being installed in systems for which it was not designed. It can deliver large quantities of water at low pressures and small quantities of water at high pressures. There is very little danger of the motor being damaged by overloading from the pump. Provided it is installed properly, there are fewer moving parts to wear out or to be replaced. The deep-well jet is most efficient when the vertical lift is between 25 and 65 feet. This type of pump will operate in wells 120 feet deep, but generally speaking is not recommended when the depth to the water is more than 85 feet. It is quiet in operation and especially suited for use in an automatic system. Neither the deep-well nor the shallow-well type need be mounted over the well casing.

Submersible Pumps Designed for wells up to 500 feet deep, the submersible-type pump can be used also in wells as shallow as 20 feet. Its great advantage is that it is a compact, integral unit with both pump and motor submerged below water level in the well casing. There is no need to have a basement or pump-

Well Depth in feet	20	40	60	80	100	120	140	160	180	200	220	240	260	280	300	320	360	400	500
PUMP MODEL	GALLONS DELIVERED PER HOUR AT 40 POUNDS PRESSURE																		
½ HP	735	690	645	595	545	...	...	...	...	...	...	...	...	...	...	...	...	...	...
¾ HP	830	790	765	725	685	655	620	570	525	...	...	...	...	...	...	...	...	...	...
1 HP	865	840	825	790	775	745	730	705	665	650	610	580	560	515	475	...	...	...	...
1½ HP	890	870	860	835	820	805	790	775	760	735	725	715	695	675	655	635	580	535	440

house installation. Only a tank and small (electrical) control box are required above ground, and they can be located conveniently in a utility room or corner of the kitchen (since there is no pump noise or vibration). Shaped like a cartridge, the integral pump and electric motor will fit any well casing of 4 inches or more in diameter. The pump is a multistage (many impellers) centrifugal type, which delivers a very dependable volume of water at relatively high pressure. The cost of a submersible type of pump, which is the most desirable, is the highest of all.

While reciprocating pumps are still in use, the jet and submersible types are by far the more popular. Most well-diggers also will install and service your pump if you do not wish to undertake the task.

Problem Solving

One of the most important considerations in owning a well is keeping it uncontaminated. Every time you open a well to clean or repair the pump, bacteria from the soil or contamination from other sources may get into the water system and cause serious trouble. Also, if a pump is left unused for long periods of time, contamination may occur. The easiest way to eliminate this hazard is to disinfect the complete system by flushing the reservoir, pump, and pipelines with a strong chlorine disinfecting solution. You can use any of the ordinary laundry bleaches, which contain about 5.25 percent chlorine.

First, measure the depth of water in your well. Then, using the following proportions, find the amount of disinfecting solution required: for dug wells, 1 to 4 feet in diameter, use 2 cups per foot of water; for drilled wells, 3 to 8 inches in diameter, use 1 cup per foot of water. Mix this with about 5 gallons of water and pour this solution into the well.

Connect one end of a hose to a convenient spigot beyond the pressure tank and let the other end discharge itself back into the well. Start the pump and let it run for about 15 minutes to thoroughly mix the disinfectant with the water standing in the well. Use the hose to flush any foreign matter from the inside of the casing and the drop pipes. Let the chlorine solution stay in the well overnight and it will kill all the bacteria that are in the water at that time. The next day, start the pump and force the treated water through the pipelines of the entire system, discharging it from the last spigot at the end of the line. After all the chlorinated water has been pumped out, let the well refill.

Wait a day or two before you have another sample tested. Do not take a sample for testing if the odor of chlorine is still present in the water.

Dug or bored wells should be disinfected in the same way as a drilled one. Lower the water level as much as possible, remove the sand, silt, and debris, and then treat with the chlorine solution. Do not try to disinfect an unprotected, unlined well, because new seepage or surface contamination will flow into the water about as fast as you can disinfect it.

If the tests are repeatedly unsatisfactory but the source of water is a clear, strong, dependable supply, you may want to install a chlorinator or some other type of water-purification equipment. Such equipment is available at a reasonable cost.

The pump and the well themselves are normally trouble-free. If your water is very hard (high in minerals) and you have a jet pump, there may be build-up in the pipes at the bottom of the well, lowering pressure. This is rare, but when it happens you will probably need a professional to remove the pipes and clean them. More common are minor annoyances, such as the vibration an above-ground pump sometimes makes when running. Bolt the pump base down using rubber washers above and below the bolt holes and rubber ferrules inside the holes. To prevent vibrations from transmitting through the house water-supply pipes, connect the supply pipe to the storage tank with a 2- or 3-foot piece of flexible plastic pipe.

Testing Your Water Supply

Most city- or utility company-supplied water does not have to be tested for bacteria. If you have a well or a spring, however, have it analyzed once each year for bacterial content—more often if the well or spring is susceptible to contamination.

Taking a Sample Special care is required when taking a water sample. Often local offices of the state board of health, university extensions, milk sanitarians, or water purifier companies will take the sample and analyze it for you gratis. There are also a number of private laboratories that will collect samples and test them for you. As a last resort, use a kit or ask the laboratory for a sterile bottle, follow their directions, and collect the sample yourself. In the absence of specific instructions, follow this procedure:

1. Obtain a sterile bottle from the testing laboratory. Let nothing except the water that is to be tested come into contact with the inside of the

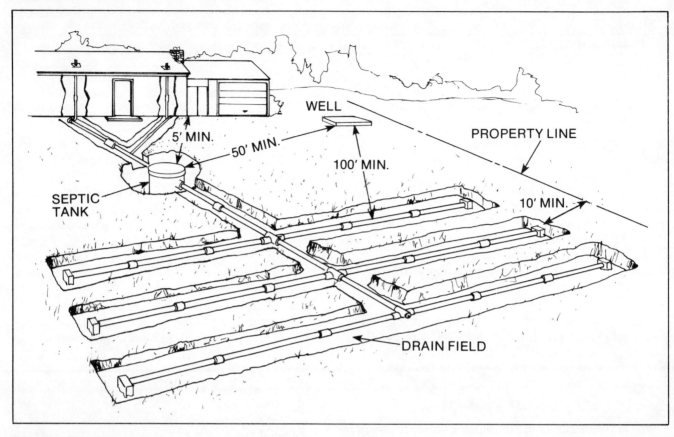

Plan and installation of a septic tank system in a yard.

bottle and cap.

2. Inspect the outside of the faucet for leaks around the handle. If there is leaking, select another faucet from which to take the sample.

3. Clean and dry the outside of the faucet.

4. Allow the water to run full force for at least a minute before you collect the sample.

5. Collect the sample, holding the bottle so that any water that comes into contact with your hand cannot run into the bottle. Cap the sample immediately.

6. Deliver the sample to the laboratory as soon as possible. A sample more than 24 hours old may not give accurate results.

No well water supply is safe to drink until it has been tested and the laboratory report indicates that it is free from contaminants. If the report is unsatisfactory, disinfect the system as described previously and retest. If it is still contaminated, remove the source of pollution (possibly a leaking septic tank uphill from the water source) and test once more. If neither of these methods work, install a chlorinator.

Troubleshooting

If a submersible pump is installed in an area where electrical storms are frequent, it is highly recommended that lightning arrestor(s) be installed. This insures proper grounding of the high voltage surges which often occur on power lines during electrical storms and prevents extensive damage to the submersible motor and cable. Most ½, ¾, and 1 HP single phase motors furnished on submersible pumps have lightning surge arrestors built into the motor. The arrestors are automatic self-contained units and do not require any special wiring or connections.

Many times the question regarding frequency of starts arises. Many factors, such as motor temperature, winding life, auxiliary components, etc., must be considered. In the selection of tank size, motor and pump size, controls, etc., you should be guided by the average number of starts per day over a period, measured in years of service.

Submersible motors will operate in accordance with the nameplate rating when the temperature of the water being pumped does not exceed 105 degrees F. Water temperature should not exceed 120 degrees F. Refer to the factory for recommendation when pumping water where the temperature may vary between 105 degrees F and 120 degrees F.

Quite often submersible pumps are installed in wells that produce less capacity than required by the pump. This situation results in: (1) an air bound pump; and (2) over-pumping the well which can be detrimental to both the pump and well. Over-pumping protective measures are as follows:

Throttling the Discharge Line After the well has been cleared of sediment and allowed to recover,

WATER PUMP TROUBLESHOOTING CHART

TROUBLE	PROBABLE CAUSE	REMEDY
Pump does not stop running.	Pressure switch is set too high.	Reset switch to lower pressure.
	Pump set too deep in well for high setting of pressure switch.	Check maximum pump setting and tank pressure. If too deep in well lower "high" pressure switch setting.
	Pump is air or gas bound.	Open disconnect switch allowing pump to stand idle for ½ to 1 hour. Start pump. See section on weak wells.
Pump starts and stops too often.	Tank water-logged.	Check valve may leak; faulty snifter valve, or clogged drain fitting. Tank may leak or improper location of drain fitting.
	Pressure switch differential set too close.	Check and reset start and stop pressures.
Low tank pressure.	Pressure switch set too low.	Reset switch for higher pressures. Check rating tables for setting and pressure.
	Pump is set too deep in well.	Check maximum allowable setting and pressure.
Pump delivers reduced capacity.	Voltage to control box or magnetic starter is low.	Check input voltage or call power company.
	Pump is too deep in well.	Check capacity and compare to rating tables.
	Pump is in sand or mud.	Pull pump. Clean strainer. Flush pump clean. Clean out well.
	Strainer or riser pipe clogged or corroded.	Pull pump. Clean strainer. Check riser pipe. Replace if necessary. Clean well.
	Well capacity low.	See section on weak wells.
Pump delivers no water. (Also see above)	Check valve installed backward.	Reverse check valve.
	Check valve stuck closed.	Free valve.
	Pump air or gas bound.	Open disconnect switch allowing pump to stand idle for ½ to 1 hour. Start pump. See section on weak wells.
Pump does not run.	Fuse blown in disconnect switch.	Replace fuse.
	Fuse size too small.	Check fuse size.
	Low voltage.	Check voltage with meter.
	Pump or motor stuck.	Check input amperage.
	Faulty disconnect switch or pressure switch.	Check control.

the discharge line should be throttled to a rate of discharge not to exceed the rate of well recovery.

1. Install a pressure gauge and shutoff valve in the discharge line.
2. Start the pump operation. As soon as the water starts to flow, close the shutoff valve tight. Record the pressure of the gauge.
3. Open the discharge valve to the approximate flow desired. Record the pressure with the water flowing.
4. Allow the pump to operate approximately 10 minutes. Shut the valve tight and record the pressure.
5. If any pressure difference is recorded between steps (4) and (2), the water level in the well has lowered. Convert pounds to feet as follows: Pressure (step 2) minus pressure (step 4) × 2.31 = ft.
6. Repeat the steps reducing flow until the level remains static.

7. The pressure of step (3), after the flow is such to allow the well level to remain static, will be the throttling pressure.
8. Pipe the horizontal run into the service tank and with the pump operating close the shutoff valve until the throttling pressure of step (3) is reached.

Low Water Safety Control Low water cut off or pump down controls are capable of taking care of variable well capacity conditions which might be experienced in some localities due to seasonal variations in the water table, etc. Probably the best control for the pump is the water level electrode type inasmuch as this control will stop the pump before it runs out of water. However, with the pump properly sized and/or throttled to the average well capacity (preferably the lowest capacity), the low water pressure safety switch or the low flow cut out safety switch will work quite satisfactorily.

Final Disposal Systems

If you do not have access to public sewage disposal, which provides the best method of sewage disposal, doubtless you want a system that can give years of trouble-free service. While such sewage-disposal systems as cesspools, privies, and chemical toilets are used, the most common one is the septic-tank system. In this system, the sewer line leads from the house to an underground septic tank in the yard. The purpose of the tank is to liquefy solid matter contained in the sewage collected in it. This is accomplished by the dissolving action of bacteria in the waste. This chemical action results in a clear liquid which flows from the tank to the disposal field. From there it disperses over a fairly large area through drain tile or perforated pipe. The tile or pipe is laid in trenches or in a seepage bed and covered with soil. The soil is planted with grass, and no part of the system is visible. For this reason, to facilitate inspection and repairs, it is good practice to keep a chart showing the exact location of the tank and other components.

Such a system should function well for many years if it is properly installed and maintained and if the soil in the disposal area is satisfactory. If the soil is not satisfactory, the sewage-disposal system will not work properly regardless of how well it was constructed and installed. Percolation tests can be helpful in determining the absorption capacity of the soil and in calculating the size of the absorption field. Most local regulations require that trained personnel, generally from local health departments, make percolation tests. These tests often are made under a wide range of soil-moisture conditions. Results are reliable only if the soil moisture is at or near field capacity when the test is made. Excessive percolation rates are obtained when there are small cracks or crevices in the soil because of insufficient moisture. False rates also are obtained when percolation tests are made in naturally wet soils that are dry during periods of low rainfall and are not thoroughly moistened before testing.

Percolation tests are made as follows:

1. Dig six or more test holes 4 to 12 inches in diameter and about as deep as you plan to make the trenches or seepage bed. Space the holes uniformly over the proposed absorption field. Roughen the sides of each hole to remove any smeared or slick surface that could interfere with water entering the soil. Remove loose dirt from the bottom of the holes and add 2 inches of sand or fine gravel to prevent sealing.

2. Pour at least 12 inches of water in each hole. Add water, as needed, to keep the water level 12 inches above the gravel for at least 4 hours, or preferably overnight during dry periods. If percolation tests are made during a dry season, the soil must be thoroughly wetted to simulate its conditions during the wettest season of the year.

Thus, the results should be the same regardless of the season.

3. If water is to remain in the test holes overnight, adjust the water level to about 6 inches above the gravel. Measure the drop in water level over a 30-minute period. Multiply that by 2 to get inches per hour. This is the percolation rate. After getting the percolation rate for all the test holes, figure the average and use that as the percolation rate.

4. If no water remains in the test holes overnight, add water to bring the depth to 6 inches. Measure the drop in water level every 30 minutes for 4 hours. Add water as often as needed to keep it at the 6-inch level. Use the drop in water level that occurs during the final 30 minutes to calculate the percolation rate.

5. In sandy soils, where water seeps rapidly, reduce the time interval between measurements to 10 minutes, and run the test for only 1 hour. Use the drop that occurs during the final 10 minutes to calculate the percolation rate.

6. Percolation tests for seepage pits are made in the same way except that each contrasting layer of soil needs to be tested. Use a weighted average of the results in figuring the size of absorption you need from the chart given here.

DATA FOR DETERMINING FIELD REQUIREMENTS FROM PERCOLATION TESTS	
Time required for water to fall 1 inch (in minutes)	Effective absorption area required in bottom of disposal trenches in square feet residences (per bedroom)
2 or less	85
3	100
4	115
5	125
10	165
15	190
30	250
60	330
Over 60	Special design using seepage pits

Note: *A minimum of 150 square feet should be provided for each individual family dwelling unit.*

Soils vary so much from place to place that it is not possible to give specific guidelines on the use of soils as absorption fields that would fit all localities. Furthermore, local health regulations vary greatly. Therefore, before you plan your sewage-disposal system, become familiar with the health regulations in your community, the permit and inspection requirements, and the penalties that may be imposed for violations. You probably also can get advice and planning aid

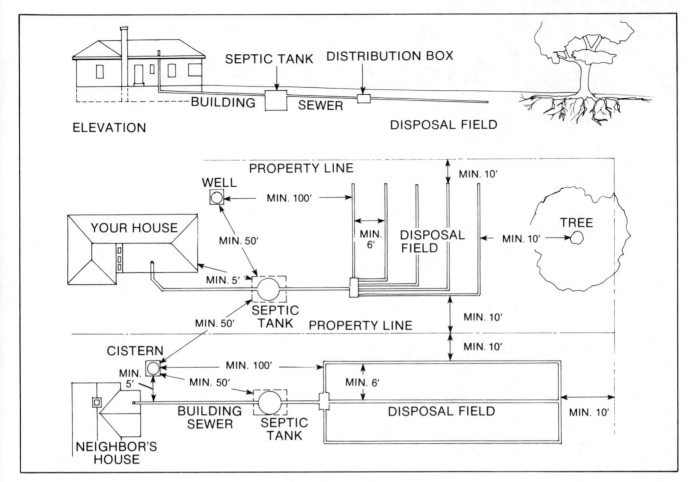

Typical minimum distance requirements applied to two common septic-tank and absorption-field systems.

from your city or county planning commission, local health department, extension specialist, engineering or agricultural department of colleges and universities, and state board of health. In addition to conforming with all local regulations, take certain precautions for your own protection and convenience in selecting the site for your sewage-absorption or -disposal field.

Installing a Septic System Septic tanks can be made of steel or concrete and usually are purchased complete and ready to be installed. The tank size—which is very important—should be sufficient for the tank to hold 100 gallons of water for each person who will live in the house, plus an additional 40 to 50 gallons per person if there is a garbage disposal. For example, the tank for a family of five must hold 500 gallons if there is no disposal and must hold 700 to 750 gallons if there is a disposal. Determine the number of persons by the number the house is built to accommodate, not the smaller number who presently may be occupying it. Two or more tanks may be joined together in sequence if necessary.

The tank may be near the house, but should—for health's sake—be at least 100 feet downhill from any well, spring, or whatever. The drawing shows the recommended minimum spacings for a tank and field. As previously explained, tank depth is determined when the house sewer is laid. Make your excavation large enough to lower the tank in conveniently and make the necessary connections—and about 18 to 24 inches deeper than the bottom of the tank level. The pit bottom must be firm and reasonably level. Line the bottom with 18 to 24 inches of gravel, crushed rock, or bricks, and level this accurately before setting the tank on it. Half-fill the tank with water and backfill around it, tamping or soaking in the earth firmly. The water in the tank and immediate area will keep it from being floated out of position by any ground-water seepage into the pit.

The house sewer run to a septic tank rarely will be more than 3 feet deep at any point. There is no hazard in hand-trenching to such shallow depth, but there may be a lot of slow, difficult digging to do. In most communities, there is someone with proper excavating machinery to speed this portion of the work, and we recommend that you contract for at least this part of the installation work, as well as the digging of the disposal field.

A house sewer line should run as straight as possible to the septic tank. Unless absolutely necessary, it should not bend, either laterally or vertically. If any bends are required, preferably they should (1) not exceed 45 degrees (one-eighth bend) in a lateral direc-

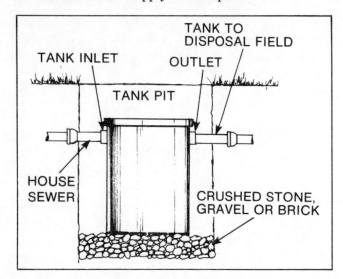

Typical septic-tank installation.

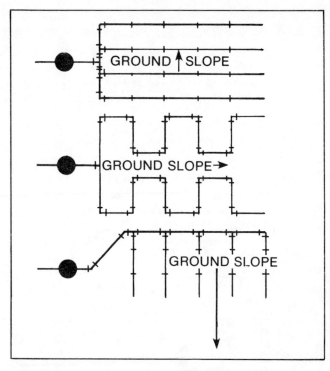

Typical disposal field patterns.

tion, and (2) not exceed 22½ degrees (one-sixteenth bend) when changing from a steeper to a shallower grade. If bends in excess of these amounts are required, they should be made with long sweep bends or several smaller bends rather than one short bend.

Whenever it is necessary to bend the sewer laterally in excess of 45 degrees, a cleanout—just ahead of the bend—is required. A cleanout is required also wherever it is necessary to reduce the grade by 22½ degrees or more. A cleanout is provided by installing a Y or a T with the vertical leg extended upward to ground level and capped—unless the sewer is deeper than 4 feet, in which case a manhole may be required. The absolute minimum pitch at any point is ⅛ inch per foot (a drop of 1¼ inches in 10 feet), and the preferred minimum pitch is ¼ inch per foot (2½ inches in 10 feet). Any pitch in excess of ¼ inch per foot is permissible. However, to avoid installation of cleanouts, it is better not to vary the pitch unless circumstances make this necessary. The last 10 feet of sewer line ahead of the septic tank must not slope more than ¼ inch per foot. A greater slope in this last 10 feet of sewer will cause water to be dumped into the tank with a violence sufficient to disrupt the chemical liquidation of waste in the tank (and result in the need for too-frequent tank cleanings). Therefore, if the tank is located so much lower than the house that the sewer must pitch sharply down to it, your only choice is to install a cleanout and lessen the sewer grade 10 feet or more ahead of the tank. In any case, whenever the tank will be more than 20 feet from the house, it is preferable to have a cleanout 5 to 10 feet ahead of the tank.

Cast-iron, fiber, underground plastic, or (if conditions permit) tile pipe may be used, depending upon local code and your preference. In most cases, fiber or plastic is recommended for all-around satisfaction and ease of installation. All watertight joints are made with these two materials by using a special solvent cement. Incidentally, distribution boxes, which often were used on the tank outlet line of older septic systems, are no longer recommended by the United States Public Health Service because they required cleaning more often than usually was given to them.

The ideal location for a disposal field is a flat, large-enough area where there are few trees or shrubs to shade the ground, and as far as possible from the house or other occupied buildings. Check your local code for requirements.

The field should be flat or nearly so because the seepage piping should be laid to pitch only about 1 inch per 10 to 12 feet and should not be buried more than 3 feet deep, or trenching might be a problem. Also, all drainage lines in the field should be separated by at least three times the width of the trenches with a minimum spacing of 6 feet. Greater spacing is desirable where possible. It is best to have all laterals of equal length to provide an even distribution of effluent (liquid sewage from septic tank). Under no conditions should a field with less than 150 square feet of effective absorption area (100 linear feet of 18-inch trench) be provided for any individual dwelling unit. The maximum length of lines should not exceed a distance of 100 feet and at least two lines of pipe should be provided. The field shape and the pattern in which the seepage pipe is laid are determined by the slope of the ground and the limitations imposed by boundary lines, buildings, and so on.

The run from the tank to the field should be watertight for a distance of at least 5 feet from the tank. Whether you make the rest of this run watertight or not depends upon where you want seepage into the soil to begin. All trenching—for this run and for the seepage pipe in the field—can be planned and dug to depth in the manner explained for a house sewer line. The pitch of the run from tank to field can be 1 inch per 5 feet or greater, but the seepage-pipe pitch, as mentioned before, should be 1 inch per 10 to 12 feet. All pipe should be buried deep enough to be protected from being crushed by whatever traffic there will be above it.

Perforated fiber or plastic pipe (laid with the holes down) is the easiest to use as seepage pipe. If tile is used, it is assembled with the loose joints covered on top with building paper or roofing material to keep dirt out. If hubless field (agricultural) tile is used, the pipes must be laid along the edge of a board buried in the ground beneath them (to ensure no sagging) with lengths spaced about ¼ inch apart and the joints covered as for tile. If the soil is light and/or sandy, no special backfilling is needed. However, if soil is heavy and nonporous, the trench should be backfilled with gravel or crushed rock to surround the pipe at least 12 inches all around.

After laying the field pipe, backfill and level the surface. Then, sod or seed it to encourage a thick crop of grass, which will hasten effluent evaporation.

SIZE AND MINIMUM SPACING REQUIREMENTS FOR DISPOSAL TRENCHES

Width of trench at bottom (inches)	Depth at trench (inches)	Effective absorption area (square feet per linear feet)	Spacing of tile lines (feet*)
12-18	18-30	1.0-1.5	7.5
18-24	18-30	1.5-2.0	8.5
24-30	18-36	2.0-2.5	9.5
30-36	24-36	2.5-3.0	10.5

A greater spacing is desirable when available area permits.

Maintenance of a Septic Tank A good septic tank normally requires no maintenance other than a yearly inspection and a cleaning about every five years. Never use disinfectants in your septic tank, because such materials destroy bacterial life—the chief agent in decomposing sewage. The tank should be cleaned when 18 to 20 inches of sludge and scum have accumulated. If a drain has not been provided, sludge may be removed by bailing or by pumping with a sludge or bilge pump. (This is usually done by a professional.) It is not necessary to remove the entire liquid contents. Sludge should be removed in the spring rather than in the fall to avoid loading the tank with undigested solids during the cold-weather months. As the sludge may contain disease-bearing bacteria, it should be disposed of by burying it, or by

any method satisfactory to the state health authorities.

Septic tank systems seldom freeze when in constant use. But if the system is to be out of service for a period of time or if exposure is severe, it may be advisable to mound over the poorly protected parts of the system with earth, hay, straw, brush, or the like.

Sump Pumps

Most pumps in plumbing are found in the water-supply system—except for the sump pump. A sump arrangement is most commonly used to eject water that seeps into a basement. The water runs into a drainage pit in the basement floor and is then pumped through a series of pipes to a safe distance from the house.

The sump pit is located below ground level. If the sump pump is installed to keep a basement dry, the pit is located at the lowest point in the floor. The size of the pit depends on pump capacity, which can vary from 2,200 gallons per hour to 4,300 gallons per hour maximum. The estimated number of gallons per hour needed determines the size of pump to buy and the size of the pit. Your local plumbing supply house will help you select the best sump pump size for your installation. There are two types of sump pumps: the upright type and the submersible type.

Upright Sump Pumps The upright model consists of a motor mounted on a pedestal. The base of the pedestal rests on the bottom of the sump pit. The motor, at the top of the pedestal, should never be under water. This is the most common and least expensive type of sump pump.

The on-off switch of an upright sump pump is controlled by the position of a ball float. When the water rises in the pit, it lifts the ball float up, causing the motor to switch on. Water is then drawn through an intake valve in the pump base and pumped up through a discharge pipe and out to the drain. As the water is pumped out of the sump pit, the float drops down and switches the motor off. The water level in the sump pit should never go lower than 6 inches or the level of the intake screen. A dry-running pump could be seriously damaged. Always check the pump if you hear it running continuously.

Submersible Sump Pump The submersible type of sump pump is usually more expensive than the upright, but cannot be damaged by flooding and requires little maintenance. The submersible type can also run for a long, continuous period without damage to the motor.

This type has two different switch mechanisms. In one type, the pressure of a certain level of water in the sump pit forces a pressure-sensitive switch to click on. After enough water is pumped out, the switch turns the pump off. These switches are easy to replace.

Other submersible pumps have a switch encased in

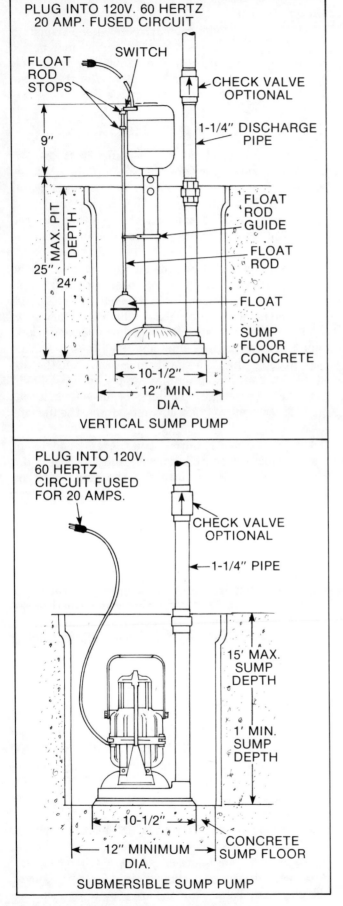

PLUG INTO 120V. 60 HERTZ
20 AMP. FUSED CIRCUIT

FLOAT
ROD
STOPS

SWITCH

CHECK VALVE
OPTIONAL

1-1/4″ DISCHARGE
PIPE

9″

MAX. PIT DEPTH

25″

24″

10-1/2″

12″ MIN.
DIA.

FLOAT
ROD
GUIDE

FLOAT
ROD

FLOAT

SUMP
FLOOR
CONCRETE

VERTICAL SUMP PUMP

PLUG INTO 120V.
60 HERTZ
CIRCUIT FUSED
FOR 20 AMPS.

CHECK VALVE
OPTIONAL

1-1/4″ PIPE

15′ MAX.
SUMP
DEPTH

1′ MIN.
SUMP
DEPTH

10-1/2″

12″ MINIMUM
DIA.

CONCRETE
SUMP FLOOR

SUBMERSIBLE SUMP PUMP

Typical sump pit and pump installations.

the tank float. If you buy this type, be sure to keep the pump and the well clean. Sand, dirt, or gravel will interfere with the pump switch and may cause it to run too long, sucking the well dry. If this happens, the pump must be repaired by a professional.

Installation Prepare the sump pit by layering it with gravel. Lower the pump into its floor crock, making sure the base is level and the switch is away from the sump wall. When the pump is oriented correctly, the discharge pipe will come up in the center of the crock.

All pumps should have a check valve. This blocks the reentry of discharged water back into the pump. Only use check valves designed for the type of sump pump you own. Usually, they will connect to variously sized discharge pipes. Before you place the pump permanently, be sure that the check valve is installed.

The cover for a submersible-type pump is one solid piece, usually made of plastic. For the upright sump pump, the cover has two holes, one for the pump shaft and one for the discharge pipe. A cover can also be made from ¾-inch exterior plywood, cut in half and bolted together to fit around the pump shaft. Install the discharge pipe to lead to a convenient drain. Use 1¼-inch to 1½-inch rigid pipe. The rigid kind is preferred to flexible because it will not bend or interfere with float switches.

This completes the installation. Make sure that everything is in place before plugging in the pump. Be sure the receptacle is grounded. If you have any trouble with the pump, turn off the power by unplugging it.

Maintenance One of the major causes of pump failure is the accumulation of sand, gravel, and dirt in the pump mechanism. To prevent this, clean the pit periodically. Use a small net or a tin can attached to a stick to catch the large debris, then stir up the remaining water and wash down the walls with a garden hose. The pump will run, clearing out the smaller debris.

Erosion of the switching mechanism is the other most common cause of pump failure. To remedy this problem, unplug the pump and replace the switch if necessary.

SUMP PUMP TROUBLESHOOTING CHART

SYMPTOM: PUMP WILL NOT RUN.

PROBABLE CAUSE	HOW TO CHECK	HOW TO CORRECT
Float switch hangs up.	Move float rod up and down to see if motor will start.	Check for switch hang-up in sump and remove cause.
No water in sump.	Same check as above.	Add water to sump.
Fuse blown.	Test power source with light or meter.	Replace fuse.
Overload in motor.	Remove plug from electric source and allow to cool for 5 minutes. After cooling for 5 minutes reconnect to power source.	Check to see if motor is submerged in liquid.
Motor has open winding.	Unplug motor cord. Take an ohmmeter and connect the 2 wires on the ohmmeter to the 2 male prongs on the motor cord. If the ohmmeter gauge fails to give any reading the motor has an open winding. If in doubt, call a licensed electrician.	Motor must be replaced.

SYMPTOM: PUMP OPERATES BUT PUMPS NO LIQUID.

PROBABLE CAUSE	HOW TO CHECK	HOW TO CORRECT
No water in sump.	Inspect sump.	Add water to sump.
Check valve installed backwards.	Look for direction of flow arrow on check valve.	Install check valve for proper flow direction.
Wrong voltage.	Check nameplate voltage against supply voltage.	Supply proper voltage supply to pump unit.
Pump inlet jammed or clogged.	Stop pump to allow liquid backflow to flush pump. If pump still fails to operate pull pump and inspect.	Remove obstruction from pump.
Leak in pipe.	Inspect sump for leaking pipe.	Replace discharge pipe.

SYMPTOM: PUMP WILL NOT SHUT OFF.

PROBABLE CAUSE	HOW TO CHECK	HOW TO CORRECT
Float hung up.	Check sump for foreign materials or improper switch located in sump.	Remove obstruction.
Discharge pipe plugged.	Check for cleanout plug or cover and remove for inspection. Be sure to disconnect electrical supply to pump.	Remove obstruction.
Discharge valve closed.	Inspect discharge line to locate discharge valve and check for open position.	Open up valve.
Liquid flowing into sump at same rate or faster rate than being pumped out.	Inspect sump or throttle inlet valve.	Install larger pump.
Discharge pipe cracked or broken off in sump.	Inspect sump for broken discharge pipe.	Replace pipe.

SYMPTOM: PUMP STARTS AND STOPS TOO OFTEN.

PROBABLE CAUSE	HOW TO CHECK	HOW TO CORRECT
Float "on and off" position set too closely.	Observe height of liquid level in sump between on and off position. If levels are close together float may be readjusted.	Readjust float rod. Be sure area for float operation is free of any obstruction which could cause float to hang up.
Sump diameter too small.	Check sump level for "on and off" levels as mentioned above.	Use larger sump or change to smaller pump.
Discharge line too long and causing backflow of liquid in sump. When pump stops or if discharged line has check valve it is not functioning properly.	View liquid in sump when pump turns off. If liquid rises when no liquid is entering the inlet, the liquid in the discharge line is flowing back into sump.	Install a check valve in discharge line. If system already has check valve replace or repair it.

SYMPTOM: FUSE BLOWS WHEN MOTOR OPERATES.

PROBABLE CAUSE	HOW TO CHECK	HOW TO CORRECT
Incorrect voltage.	Check voltage on motor nameplate. Make sure it is same as power source.	Change source of supply to correct voltage.
Fuse too small.	Check instructions for proper fuse size.	Replace with new fuse.
Motor shorted or has locked rotor.	Check amperage draw with amprobe.	Pull unit and check for jammed impeller. If impeller is free it will be necessary to replace motor.
Switch or switch cord shorted.	Remove switch from system by plugging motor directly into power source. If system operates without switch, problem is in switch assembly.	Replace switch assembly.

Possible Problems With Your Plumbing System

As was mentioned in earlier chapters, a plumbing system is built from materials that should last decades. Also, since most of that system is buried inside house walls and thus does not get much wear and tear, routine maintenance and a few simple repairs can keep it in top working condition. The average home owner will find the following relatively simple procedures for taking care of clogged drains, leaky faucets and pipes, and faulty toilets well within his or her abilities.

Clogged Drains

Drains may become clogged with objects dropped into them or by accumulations of grease, dirt, or other matter. If only one fixture is clogged, the problem can usually be solved by clearing the trap or waste connection. However, if several fixtures are clogged, chances are that the blockage is in the branch drain or its connections to the sewer or septic tank.

Fortunately, most single fixture clogged drains are easily removed. For example, some clogs can be cured by merely removing debris—hair, string, paper, and the like—on a strainer or a drain stopper. But with some guidance, most home owners can take on even more complicated blockage jobs. Let us first look at the simple procedures for freeing a kitchen sink, lavatory, tub, shower, or floor drain.

To clean out the grease and hair caught in the stopper, remove the strainer or stopper and clean it. Most strainers must be rotated before they will lift out. If the strainer is unremovable, clean it with a toothbrush saturated with grease-cutting detergent.

The method of removing the stopper depends on the style of the fixture. Some lift out; others twist one-quarter turn, then lift out. If the stopper cannot be removed by rotating it, it is held by a pivot rod and retaining nut under the sink. Put a pail under the sink drain and then loosen the assembly coupling underneath the basin. Pull the pivot rod back to release the stopper.

If the stopper is clean and the drain is still blocked, try a force cup plunger. After the strainer or stopper has been removed, spread newspaper underneath the sink. Fill the sink about one-quarter full. Plug the overflow with wet rags; when working on a double sink, seal the other drain and overflow with rags, too.

To obtain the necessary suction, roll the force cup into the water so that as little air as possible is trapped within the cup, then center the cup over the drain. A coating of petroleum jelly around the lip of the plunger makes a tighter seal. Pump vigorously 10 to 20 times, then jerk the plunger up, allowing the water to rush down the drain. If the drain is still blocked, plunge it again, repeating the procedure two or three more times. If you can sense the rhythm, you can send drain-clearing jets of water down with each plunge.

If plunging will not start a run-off, the next step is to drain the trap. Some traps have cleanout plugs, while others must be removed entirely. Some older bathtub and shower installations are equipped with a cylindrically shaped drum trap located at floor level beside the fixture which has a cover that must be removed to get at the drain. As soon as you remove the cleanout plug or the trap cover, water could gush out, so position a bucket under the trap to catch any overflow. Also, make sure the faucets are turned off. If the trap must be removed, use an adjustable wrench and wrap the jaws with tape or cloth to protect the chrome finish on the trap fitting.

Clear any blockage in the trap, then clean the trap completely with a bottle brush and soapy water. If there is no blockage in the trap itself, probe the pipes with a bent coathanger to locate possible debris. When none of these methods work, you have to resort to more sophisticated plumber's tools.

One of these is a plumber's auger, a snake-like coiled wire, with one end barbed and one end usually enclosed by a metal sleeve with a crank. But, it may be just a coiled cable. A 10-foot drain auger is preferable, because it can reach through most drain pipes to the main stack. Work the auger around corners by turning the crank. Back it in and out, using sharp thrusts, and cranking it to probe through the obstruction. When you contact the blockage, keep turning the crank in the same direction while withdrawing the obstruction slowly. In this way, the auger will continue to engage the blockage, whereas an opposite turn would release it. The auger can also be used to free tub blockage.

If you do not have an auger, you can improvise with a garden hose, with or without water pressure, to clear a drain. Snake the hose through the drain until it reaches the stoppage. Jab and twist, trying to loosen it. Turn on the water and use the water pressure to blow out the blockage. While a force cup or auger can be used to free floor drain blockage, the hose method usually works best.

Clogged Toilets Most of the time a toilet can be cleared with a force cup plunger. If the toilet bowl is full, empty it halfway before plunging. If possible, use a special toilet plunger with a foldout rim that fits into the drain.

If a few rounds of plunging do not clear the blockage, try using a closet or toilet auger, which has a handle especially designed to guide the auger into the toilet trap. Of course, you can use a regular drain auger if you are willing to get your hands wet.

Insert the auger and turn the crank, guiding it around sharp turns in the bowl. Try hooking the blockage or breaking it up rather than jabbing it or pushing it in deeper. If none of these methods work, it may be necessary to remove the bowl to remove the obstruction.

Chemical Cleaners In some instances, chemical cleaners may be used to free a clogged drain. But, as a rule, these cleaners do not dissolve the blockage completely, and you then have to open the trap anyhow. Furthermore, using drain cleaners once a week to keep drains clear, as manufacturers suggest, pollutes the environment.

If in exceptional cases you do decide to use chemical cleaners, heed these guidelines:

1. Wear rubber gloves and goggles.
2. Never plunge after a drain cleaner has been used.
3. Read all cleaner labels and match the type of cleaner with the kind of blockage. In general, use alkalies (i.e., lye) in the kitchen sink to cut grease, and use acids in bathroom fixtures and floor drains. Never mix alkali and acid cleaners. Read the warning labels and match the cleaners to the sink and pipe material. For example, some cleaners must not be used on stainless steel sinks.
4. Do not use drain cleaners in the toilet bowl.

Unclogging Main Drains If these methods fail to locate and remove the blockage, the main stack or the sewer lines may be clogged, especially if more than one fixture is blocked. First, determine where the clog is by checking other fixtures in the house. You can usually

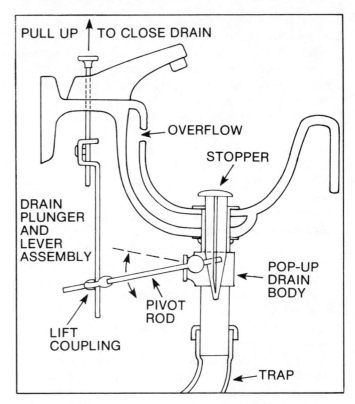

How a typical lavatory faucet with pop-up drain operates. The left coupling attached to the pivot rod must be loosened to remove the stopper.

tell when the main stack is clogged because, logically, other fixtures back up also. For example, if toilets on the first and second floors back up at the same time, you know the blockage is not in the fixture drains themselves but in the common drain pipe.

Most newer plumbing systems have cleanout plugs all along the stack. Beginning with the plug closest to the lowest clogged fixture, loosen the cap slowly. Have a large bucket or an empty garbage can on hand to catch the overflow. If no or very little water comes out, you know that the blockage is between the plug and the fixture. If there is water present, the blockage is between the plug and the sewer line. When the line has drained, remove the plug completely and insert an auger. When replacing the plug, clean off the corroded

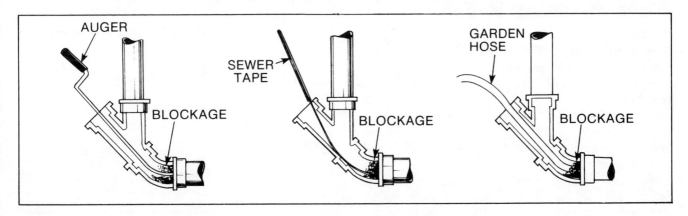

Methods of unclogging a sink drain.

threads and apply grease.

Sometimes, an old plumbing system does not include cleanout plugs. You can still clear the main stack of such a system through the roof vent, but you need a drain and trap auger long enough to reach the blockage. Feed the auger down the stack, turning it as already described. Once the obstruction is broken through, flush out the stack with a garden hose.

If the system is still clogged, tree roots could be blocking the sewer lines. Installed properly, drainage pipes and sewer lines are root-proof. However, if the plumber used inferior materials or did a shoddy job, roots could grow in at the joints between the pipe. Relentless trees will extend their roots 60 feet to seek water. The roots of willows reach up to 150 feet!

A power auger—rented or operated by a plumber—can cut the roots out but generally the only lasting solution is to dig up the damaged section and repair or replace it.

Preventing Blocked Drains Drains seem to clog just at the wrong time: when you are ready for bed, when you're late for work, or just when guests are knocking at the door. But, regular maintenance will prevent these emergencies.

Grease, hair, food particles, toilet tissue, and other debris can build up in drain pipes. Do not throw anything but toilet tissue down a toilet. Also, keep toilet seats closed to help keep jewelry and rings out. When family members clean their hairbrushes, they should put hair in the wastebasket, not down the drain.

Never put any food down a kitchen sink drain unless there is a garbage disposal unit installed. Keep strainers in and just cock them to one side to allow water to drain out of the sink. If you must pour grease down the sink, follow it with boiling, sudsy water to keep the drains clear.

Recurring problems can also be due to drain structure. Drain pipes could be too large or too small, not angled enough or too much—all of these conditions create clogging. Cheap construction leaves sealer between pipe joints sticking into the passage to catch the debris. Seek professional advice.

Dripping Faucets

A leaking faucet is not only annoying, it is extremely wasteful. Twenty-five drips per minute wastes a bathtub full of water per week. But, do not spend the money you will save fixing a drip by calling in the plumber; you can usually stop the leak yourself. As mentioned earlier in the book, there are five basic types of faucets: washer-type compression faucet; diaphragm faucet; cartridge faucet; single lever ball; and single lever disk.

Washer-Type Faucets With most washer-type faucets, each unit has its own spout, but many modern ones have two spigots—hot and cold—sharing one spout. Since such modern fixtures are in reality two

entirely separate faucets, you may have to spend a few minutes determining whether the hot or the cold spigot—or both—is leaking.

The washer, of course, is a vital part of the faucet. When it wears out, it does not seal the valve seat tightly. Never force the faucet closed to stop a drip. Instead, replace the old washer or packing, or dress the valve seal.

Follow these steps to replace a washer:

1. Shut off the water supply and open the faucets to drain the line. Next, put the stopper or old rags in the drain so small parts do not slip down.
2. It may be necessary to remove a decorative cap in the center of the handle. If it is the press-on type, pry it up with a thin-bladed screwdriver or table knife. If it is threaded, wrap the edge with tape and use a monkey wrench. Remove the handle screw and pull off the handle.
3. Remove the escutcheon and unscrew the packing nut. Turn the stem in the on direction to remove it.
4. Remove the small screw that holds the worn-out washer to the end of the stem. Remove the washer, prying it away with the screwdriver, if necessary. Replace it with a faucet washer that will tolerate both hot and cold water. Clean off all corroded parts before reassembling the faucet.

Sometimes a washer will not fit the stem exactly. Select the next larger size and trim it down by pressing it onto a drill bit, then chucking the bit into an electric drill. Run the drill on low speed and lightly touch the washer to a strip of abrasive paper taped down to a table. Compare it to the old washer. Do not use a washer that is too small. If you are stuck without the right size replacement washer, a temporary solution is to turn the old washer upside down and screw it back onto the stem.

While the faucet is disassembled, check the faucet

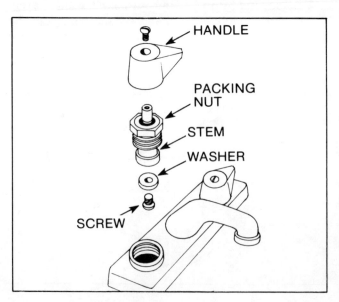

Typical washer-type faucet.

seat by feeling it with your fingertip. It must be clean and unmarred for a tight seal. If it feels smooth, fine. If not, use a rag on a toothbrush handle to remove the scale and grit. If you feel nicks or notches, use a seat dressing tool.

If the seat cannot be ground smooth, replace it with a new one. You will need a seat faucet wrench to install it. Some valve seats are permanently attached. In that case, press a slip-on seat over the old seat, after cleaning off the rust or dirt.

To replace washers in some types of bath faucets, you will need an extended socket wrench of the right size. First, remove the handle and escutcheon. If there is a sleeve over the stem, protect it with tape and unscrew it. Second, fit the socket wrench over the bonnet and remove it. (The bonnet and stem will come out of the body together.) Third, replace the washer and reassemble.

If water leaks out around the base of a stem faucet, try tightening the packing nut. If leaking persists, the packing is probably worn. Packing is made from a graphite impregnated material and is used for sealing faucet stems. Self-forming packing looks like black, shiny twine that is twisted around the stem, underneath the packing nut. Use one and a half times the amount of new packing needed to fill the packing

nut—it will be compressed—and seal tightly. In more modern faucets, self-forming packing has been replaced by packing washers or O-rings.

Cartridge Faucets   In a cartridge faucet, the stem is replaced by a cartridge. To close off the water, the cartridge presses against a spring-based washer inside the faucet body. Often, all you must do to stop a drip is replace the washer-spring unit.

Remove the cartridge faucet as you would a stem faucet. Make note of how the keys on either side of the cartridge align with keyway slots on the body of the faucet. Then, using long-nosed pliers, pluck out the rubber washer and the metal spring. If the spout still drips after the faucet is reassembled, you will have to replace the entire cartridge.

Diaphragm Faucets   Diaphragm washers usually last longer than the standard washers on stem faucets. The diaphragm functions as the washer and O-ring. Some contain packing or an O-ring anyway behind the diaphragm. To repair, remove the handle and pull out the spindle assembly. Pull the diaphragm off the end of the stem and press on a new one.

Single Lever Faucets   Single lever faucets combine hot- and cold-water lines into one spout. Although there are many different designs, underneath the chrome there are two basic types: disk valve faucets and ball valve faucets.

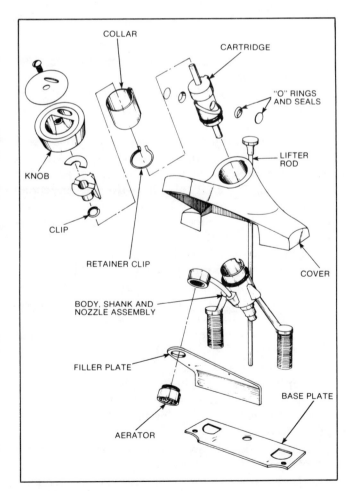

Repairing a cartridge faucet.

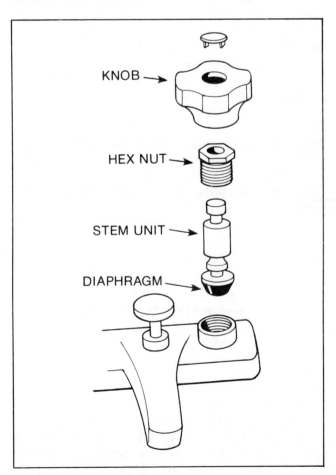

Repairing a diaphragm faucet.

The disk valve faucet mixes hot and cold water through two disks. The top disk slides when the handle is turned, while the bottom disk remains stationary. The disks are usually made of ceramic, glass, or plastic with two or three openings. When the holes are aligned, the water flows through. When they are not aligned, the faucet is off.

Disks rarely wear out, but sometimes dirt can get between them. Clean the disks, and smear petroleum jelly between them. Be sure not to get petroleum jelly on any rubber parts.

The base of the lever of the ball valve faucet is round. Inside is a ball with three holes through which the water flows, depending on their positions in relation to two other holes in the faucet body.

When these types of faucets leak around the stem or lever, try tightening the compression nut. This type of faucet also has a circular packing ring or O-ring on top of the ball that may be worn.

If the faucet drips, spring-based valve seats in the faucet body may need replacement. To get to them, lift up the handle and remove the screw underneath it. Unscrew the cap, then lift out the whole assembly.

Each model of single-handle shower-tub control is slightly different, and so is its repair. For this reason, it is necessary to consult the manufacturer's instruction sheet. Because replacement parts are not interchangeable, any items for the repair job must be purchased from your plumbing supply house or distributor handling that specific brand. If they cannot assist you, write to the manufacturer to give you some idea as to the problems that can arise.

Aerators in faucets can restrict the flow of water when clogged. If the aerator is only slightly plugged, it sometimes is possible to clear it by removing it from the faucet and holding it under the stream from the faucet, upside down. This may flush out the tiny particles of rust and other debris that plug the narrow passages inside the aerator. If this "back-flushing" does not clear the aerator, then it must be disassembled for cleaning. Servicing is simply a matter of taking them apart, keeping the parts in order, cleaning them, and reassembling.

Leaky Valves and Supply Stops

As a rule, repairing a valve that has seen long service is not too practical and, in these cases, replacement is the answer. This is especially true of gate valves and globe valves.

If you have a compression stop that will not shut off, replace the washer. First, remove the old washer and scrape away any residue. Install the new washer, making sure to tighten the screw securely. (The washers are the same style as those used in faucets.) If the valve continues to leak, dress the valve seat with a reseating

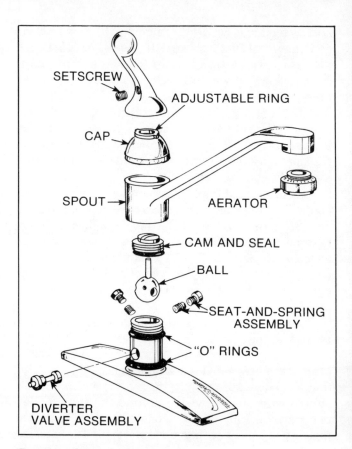

Repairing a single lever ball faucet.

tool. (The same tool as is used on faucets.) Use it lightly to prevent damaging the surface, and be sure to remove any chips before assembling.

One common household valve is different from the rest. Called a diverter valve, it is found in a tub-shower combination and in a sink with a hand sprayer. The diverter valve is inside the spout in a tub. A clockwise turn moves the stem into the valve seat, closing the pipe to the tub spout and forcing the water through a hollow housing, usually plastic, to the shower head. A counterclockwise turn of the handle pulls the stem back and opens the pipe to the tub spout. If the valve leaks, remove the tub spout and disassemble it the same way you would a washer stem faucet. Replace the worn washers, packing, or badly worn metal parts.

If water is not reaching the shower head, first check for clogging or insufficient water pressure. As a last resort, unscrew the tub spout and inspect the valve. If the plastic housing is at fault, the valve is irreparable. Since diverter valves are built in, you usually have to get a new tub spout.

Supply Stops Supply stops are valves which control water lines. If they leak, shut off the water, then remove the old valve. To install a new supply stop, proceed as follows:

1. Apply pipe compound or Teflon tape to the threads of the supply pipe.
2. Screw the lavatory stop onto the supply pipe. Tighten it with a wrench. (Protect the chrome

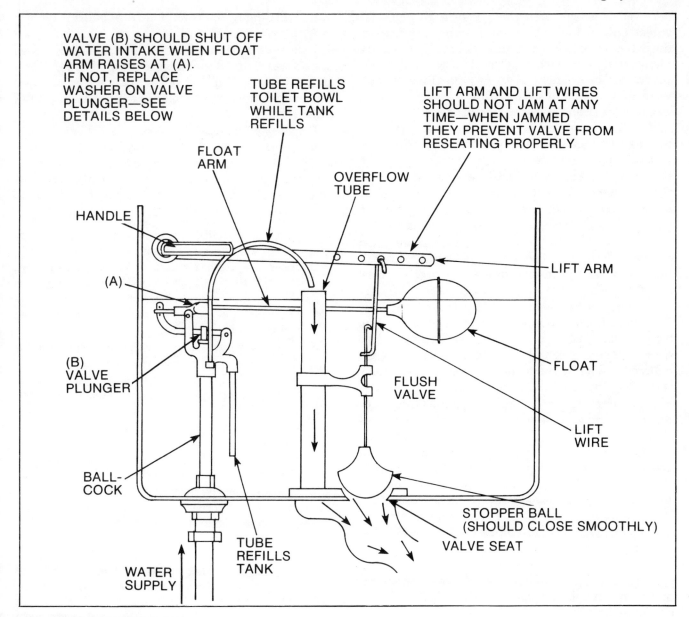

VALVE (B) SHOULD SHUT OFF
WATER INTAKE WHEN FLOAT
ARM RAISES AT (A).
IF NOT, REPLACE
WASHER ON VALVE
PLUNGER—SEE
DETAILS BELOW

TUBE REFILLS
TOILET BOWL
WHILE TANK
REFILLS

LIFT ARM AND LIFT WIRES
SHOULD NOT JAM AT ANY
TIME—WHEN JAMMED
THEY PREVENT VALVE FROM
RESEATING PROPERLY

FLOAT
ARM

OVERFLOW
TUBE

HANDLE

LIFT ARM

(A)

(B)
VALVE
PLUNGER

FLUSH
VALVE

FLOAT

LIFT
WIRE

BALL-
COCK

STOPPER BALL
(SHOULD CLOSE SMOOTHLY)

VALVE SEAT

WATER
SUPPLY

TUBE
REFILLS
TANK

How a tank toilet operates.

with friction tape.)

3. Align the outlet end of the lavatory stop with the faucet shanks. Remove the nut and compression sleeve from the outlet end of the lavatory stop and install it on the flexible supply tube.

4. Insert the end of the flexible supply tube into the outlet end of the lavatory stop and tighten the compression nut (D).

Faulty Toilets

The best beginner's instruction on toilet repair is to flush your own and see how it works. A tank-type toilet flushes by the pressure of the water rushing from the tank down into the bowl. Tankless flush-type toilets have no tanks and flush with the direct pressure of water lines entering from above.

Tank Toilets To understand a tank toilet and

where it can develop problems, imagine a bottle with a stopper. If you invert the bottle and pull out the stopper, water cascades out with considerable pressure. This is the pressure which flushes the tank.

In the tank, the stopper is on the inside, and every time you push down on the handle, a wire or chain lifts the stopper up and all of the water (10 gallons or so) passes into the bowl. The stopper is called a tank ball, and the opening in the tank is the outlet valve.

During flushing, the tank ball remains suspended above the valve because it is hollow and will float. However, once all the water is gone, it drops down and seals the outlet so that the tank can refill.

The seal between the tank ball and the valve seat must be tight or the toilet will constantly leak water— not much, but enough to cost you money. The seat could be dirty or nicked, or the ball could be misshapen or misaligned.

To repair a leaky valve, make sure the tank ball guide is aligned exactly with the valve seat because the ball must drip squarely into the seat. If necessary, realign or replace the ball. If this does not fix the leak, shut off the water, then flush the toilet so that no water remains in the tank. Clean off the valve seat with a sponge or cloth and press the handle a few times. If the tank ball does not seal because it has deteriorated or become soft or misshapen, replace the ball. If the valve seat is chipped, replace it.

If the toilet flushes completely only when you hold the flush handle down, the lift wire is too long and the ball or valve cannot rise high enough to clear the outrushing water. Shorten the wire.

If the tank fills but the toilet keeps on singing and you have checked or replaced the valve and tank ball, the water may be running through the overflow tube into the bowl.

The overflow tube channels water into the trap after flushing is finished. Once the tank has filled up, the water should also shut off to the overflow tube. If it does not, there is usually something wrong with the ballcock assembly, the toilet mechanism that brings water into the tank.

The ballcock assembly consists of an inlet tube (ballcock), and inlet valve, a refill tube, and a float ball attached to a float arm. When the toilet tank empties, the float ball drops toward the bottom of the tank and opens the inlet valve. Water rushes into the tank through the refill tube and into the toilet trap through the overflow tube. The hollow float ball floats on top of the tank water as it rises; when it reaches a predetermined level, the inlet valve shuts off. All flow through the refill and overflow tubes should stop. There should be no bubbling or gurgling noises.

If the float ball is leaking, it will lie near the bottom of the tank, unable to trigger the inlet valve to turn the water off. Simply unscrew the old float ball and replace it with a new one.

If the float ball is doing its job, lift the float arm up as high as it will go. If the water shuts off, bend the float arm down ¼ to ½ inch and test. If the water still does not shut off, the trouble is often a malfunctioning inlet valve.

Remove and replace the washer in the flush valve by first unscrewing the float arm. Lift the valve plunger out and replace the washer. Again, as in a faucet, be sure that the washer fits exactly. In many cases, it is best to replace the entire flush valve assembly, which is done as follows:

1. Remove the old flush valve along with the lift wires, washer, and locknut.
2. Insert the discharge tube of the new flush valve through the tank bottom with the large rubber cone washer in place. Turn the unit to locate the overflow tube properly before tightening the locknut under the tank.
3. Center the guide arm over the valve seat. Install the lift wires; then screw the flush ball to the lift wire assembly. Check to be sure that the flush ball will drop into the exact center of the flush valve seat.
4. To complete the job, replace the refill tube above the overflow tube.

To replace the entire ballcock assembly, work with the water off and the tank empty. First, unscrew the slip joint nut from the threads protruding underneath the tank. This disconnects the water line, so be sure the water is off. Then, remove the large nut directly above the slip nut joint. The valve pipe and the valve may now be lifted out of the tank. Mount the replacement valve by reversing the steps. Place the shank of the new ballcock assembly through the opening. There should be a large, rubber, cone-shaped washer in place. Tighten the nut, then slip one end of the refill tube over the plastic lug at the top of the ballcock. Place the plastic holder over the opening of the refill tube on the closet flush valve. Position the refill tube in the holder, then install the float rod and the float.

Sealing the Toilet Another place a toilet can leak is between the base of the bowl and the floor. To find out exactly where your leaks are occurring, add blue dye to the toilet tank at night, then first thing in the morning check for dye at the base and inside the bowl.

If there is a leak between the toilet bowl and the floor, the cause is usually worn-out putty seal. Plumbers used to use putty to seal toilets, but after a while, the putty disintegrates.

Unbolt the bowl from the floor after turning off the water. Turn it upside down and, using a knife or screwdriver, pry out all of the old putty and plaster. Replace it with a rubber or wax ring that will seal the toilet almost indefinitely. Reposition the toilet, and seal around the edge of the bowl with putty.

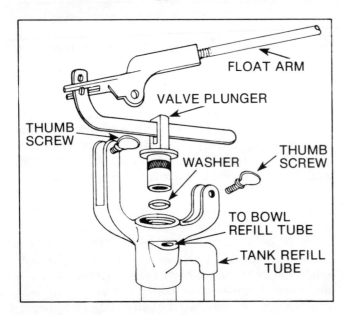

Details of a typical ballcock valve.

Tank Condensation The cold water in a toilet tank will condense moisture out of warm summer air. Constant dripping can mildew tiles and rugs and corrode metal parts on the outside of the tank.

Condensation can usually be corrected with a waterproof insulation jacket found at plumbing supply stores. You can make one from ½-inch foam rubber cut to size and glued to the inside of the toilet.

Flush Valve Toilets A flush valve toilet does not have a tank to store water. Instead, it is connected to a large pipe, usually 1 to 1½ inches in diameter, providing a speed of 30 gallons per minute. If the toilet functions well, this type of toilet saves water. If it malfunctions, however, flooding or great loss of water is the result. The valves are either piston- or diaphragm-operated and usually last a long time without service.

There are three parts to the flush valve mechanism: the emergency shutoff valve, the handle, and the pressure valve itself. Most handles and stop valves are similar, but there are two types of pressure valves, a diaphragm- and a piston-operated valve. The diaphragm valve has a rounded cover larger than the pipe that leads to it. The piston valve has a cover about the same size as the pipe. Fixing both of these types of valves usually means replacing worn parts. Check the specific manufacturer for complete instructions.

Usually, flush valve toilets have an adjustment screw to regulate water volume. On the diaphragm-type toilet, the shutoff valve regulates the flow. Open the valve one or two turns so that the toilet flushes completely but refills the bowl. The piston-type toilet has a regulating screw beneath the cap. On this type, leave the shutoff valve open completely and adjust the regulating screw that is beneath the cap. Turn the screw clockwise, then experiment by turning back one or two turns for a longer flush cycle and forward for a shorter one.

Water Conservation in Your Toilet Tank You can buy water-saving and water-efficient toilets, but it may be cheaper to adjust or modify your own toilet to save water.

The first way to save water is to check for leaks and tighten valves. Use the blue dye check described. Also, bend down the float arm so that the inlet valve will shut off sooner, thus saving some water.

Most people have heard of putting a brick into the tank to displace water. This is not a good idea because a brick will eventually disintegrate and clog the valve seat or the plumbing, eventually creating more leaks. Instead, use a 1-gallon plastic bottle and cut off the top so it will fit into the tank. Put a few stones into the bottle to weight it down. Fill the bottle with water and sink it to the bottom of the tank so that it will not interfere with the flushing mechanism. This will reduce the water by about 1 gallon with each flush. If this works with one bottle, try to sink another bottle into the tank. If there is no room for bottles, buy a commercial dam which reduces water usage without using so much room.

You can create a water-saving dual flush toilet if your tank uses the plunger-type inlet valve.

Turn off the water supply to the tank and flush the toilet. You want to weight the inlet valve so just enough water leaves the tank before the valve closes. Apply bits of solder until you have reached that point. To flush liquid waste with just the water you need and no more, press the handle down and release; for solid waste, hold the lever down to flush.

Once you have installed or modified the toilet, be sure that the water returns to the proper level. The trap must be full so that sewer gas does not enter the house. Mark the level inside the bowl with a grease pencil before you begin to work.

Another way to check is to add toilet cleaner to the water in the tank, then flush. If the water in the bowl is clear or even slightly blue after you flush it, fine, but if it is a deep blue, increase the flush power of the tank.

With the flush valve type of toilet, whether diaphragm or piston type, use the regulating screw to adjust the amount of water used.

Leaky Pipes

Leaks in pipes usually result from corrosion or from damage to the pipe. Pipes may be damaged by freezing, vibration caused by machinery operating nearby, water hammer, or by animals bumping into the pipe.

Occasionally, waters are encountered that corrode metal pipe and tubing. (Some acid soils also corrode metal pipe and tubing.)

The corrosion usually occurs, in varying degrees, along the entire length of pipe rather than at some particular point. An exception would be where dissimilar metals, such as copper and steel, are joined.

Treatment of the water may solve the problem of corrosion. Otherwise, you may have to replace the piping with a type made of material that will be less subject to the corrosive action of the water.

It is good practice to get a chemical analysis of the water before selecting materials for a plumbing system. Your state college or health department may be equipped to make an analysis; if not, you can have it done by a private laboratory.

Repairing Leaks Pipes that are split by hard freezing must be replaced. A leak at a threaded connection can often be stopped by unscrewing the fitting and applying a pipe joint compound that will seal the joint when the fitting is screwed back together.

Small leaks in a pipe can often be repaired with a rubber patch and metal clamp or sleeve. This must be considered as an emergency repair job and should be followed by permanent repair as soon as practicable.

To seal leaks in plastic pipes, you can often use

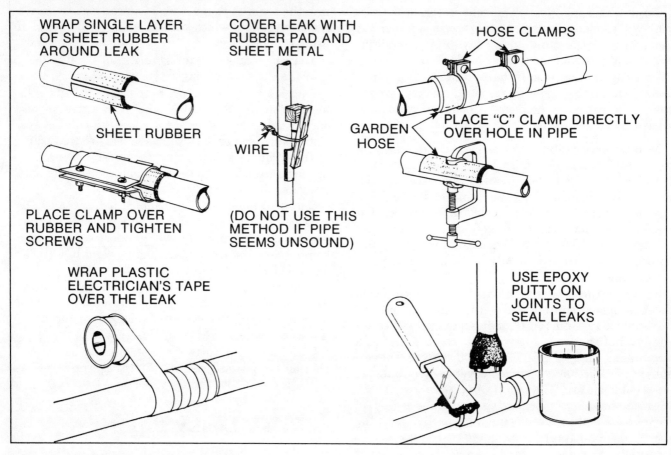

WRAP SINGLE LAYER
OF SHEET RUBBER
AROUND LEAK

SHEET RUBBER

PLACE CLAMP OVER
RUBBER AND TIGHTEN
SCREWS

COVER LEAK WITH
RUBBER PAD AND
SHEET METAL

WIRE

(DO NOT USE THIS
METHOD IF PIPE
SEEMS UNSOUND)

HOSE CLAMPS

PLACE "C" CLAMP DIRECTLY
OVER HOLE IN PIPE

GARDEN
HOSE

WRAP PLASTIC
ELECTRICIAN'S TAPE
OVER THE LEAK

USE EPOXY
PUTTY ON
JOINTS TO
SEAL LEAKS

Good temporary repairs for a leaking pipe.

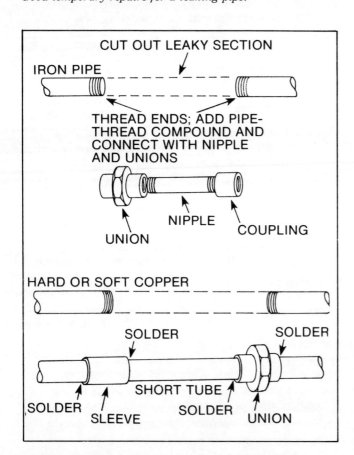

CUT OUT LEAKY SECTION

IRON PIPE

THREAD ENDS; ADD PIPE-
THREAD COMPOUND AND
CONNECT WITH NIPPLE
AND UNIONS

NIPPLE

UNION

COUPLING

HARD OR SOFT COPPER

SOLDER

SOLDER

SOLDER

SHORT TUBE

SLEEVE

SOLDER

UNION

Method of making a permanent repair for a leaking pipe.

plastic electrician's tape, wrapping it around tightly and using an ample amount of tape on both sides of the hole. Then, have the damaged section replaced.

Epoxy putty is a good joint sealer for a leak around the tee and ell joints of cast-iron pipes. Spread it on thick. Before you do any patching, turn off the water at the main entrance, open the taps, and drain the line.

Large leaks in a pipe may require cutting out the damaged section and installing a new piece of pipe. At least one union will be required unless the leak is near the end of the pipe. You can make a temporary repair with plastic or rubber tubing. The tubing must be strong enough to withstand the normal water pressure in the pipe. It should be slipped over the open ends of the piping and fastened with pipe clamps or several turns of wire.

Vibration sometimes breaks solder joints in copper tubing, causing leaks. If the joint is accessible, clean and resolder it. The tubing must be dry before it can be heated to soldering temperature. Leaks in places not readily accessible usually require the services of a plumber and sometimes of both a plumber and a carpenter.

Tank Leaks Leaks in tanks are usually caused by corrosion. Sometimes, a safety valve may fail to open and the pressure which has developed will cause the tank to spring a leak. While a leak may occur at only one place in the tank wall, the wall may also be corroded thin in other places. Therefore, any repair should be considered as temporary, and the tank should be replaced as soon as possible.

A leak can be temporarily repaired with a toggle bolt, rubber gasket, and brass washer. You may have to drill or ream the hole larger to insert the toggle bolt. Draw the bolt up tight to compress the rubber gasket against the tank wall.

Frozen Water Pipes

In cold weather, water may freeze in underground pipes laid above the frost line or in pipes in unheated buildings, in open crawl spaces under buildings, or in outside walls.

When water freezes it expands. Unless a pipe can also expand, it may rupture when the water freezes. Iron pipe and steel pipe do not expand appreciably. Copper pipe will stretch some, but does not resume its original dimensions when thawed out; repeated freezings will cause it to fail eventually. Flexible plastic tubing can stand repeated freezings, but it is good practice to prevent it from freezing.

Preventing Freezing Pipes may be insulated to prevent freezing, but this is not a completely dependable method. Insulation does not stop the loss of heat from the pipe—it merely slows it down—and the water may freeze if it stands in the pipe long enough at below-freezing temperatures. Also if the insulation becomes wet, it may lose its effectiveness.

Electric heating cable can be used to prevent pipes from freezing. The cable should be wrapped around the pipe and covered with the proper insulation.

Thawing Use of electric heating cable is a good method of thawing frozen pipe, because the entire heated length of the pipe is thawed at one time. Thawing pipe with a propane torch can be dangerous. The water may get hot enough at the point where the torch is applied to generate sufficient steam under pressure to rupture the pipe. Steam from the break could severely scald you.

Thawing pipe with hot water is safer than thawing with a propane torch. One method is to cover the pipe with rags and then pour the hot water over the rags (B).

If these methods do not prove successful, remove a part of the pipe (at a union), insert a small pipe as far as it will go, and pour in boiling water (C), allowing the returned water to flow into a bucket. (A length of rubber tubing, instead of pipe, will also work satisfactorily.)

When thawing pipe with a blowtorch, hot water, or similar methods, open a faucet and start thawing at that point. The open faucet will permit steam to escape, thus reducing the chance of the buildup of dangerous pressure. Do not allow the steam to condense and refreeze before it reaches the faucet.

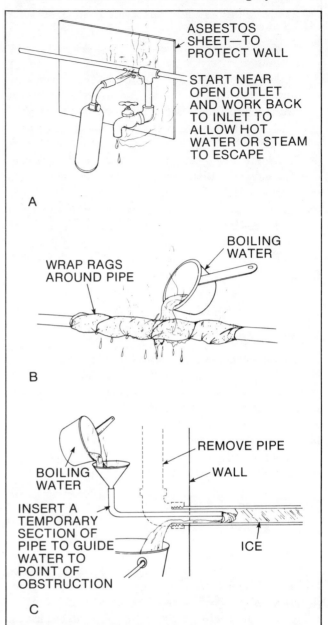

Methods of thawing a frozen pipe.

Underground metal pipes can be thawed by passing a low-voltage electric current through them. The current will heat the entire length of pipe through which it passes. Both ends of the pipe must be open to prevent the buildup of steam pressure. Caution: This method of thawing frozen pipe can be dangerous and it should be done by an experienced person only. It cannot be used to thaw plastic tubing or other non-electricity-conducting pipe or tubing.

Draining the Supply System If you are closing a house for a period of time—especially over the winter—you should drain the water-supply system. Even a small amount of water in a valve can cause damage if it freezes. Do each of the following in order:

1. Shut off the main supply stop.
2. Close the main shutoff valve inside the house.

This valve normally has a small drain cock for draining the valve and connecting pipes. Do not open this drain cock until the system has been emptied.

3. Open all faucets and flush all toilets, working from the top floor down. Be sure the hot-water heater is off and then drain the hot-water tank.

4. After all faucets have stopped draining, open the drain cock on the main stop valve, and allow the water to flow into buckets. If there is no main stop valve, disconnect a section of the lowest point in the system.

5. If the house is supplied by a pump, you must remove the water from the pump, tank, and connecting lines.

When leaving a house unoccupied, the traps for each fixture must also be drained.

Water Hammer and Noises in Pipes

Water hammer sometimes occurs when a faucet is suddenly closed. When the flow of water is suddenly stopped, its kinetic energy is expended against the walls of the piping. This causes the piping to vibrate, and leaks or other damage may result. To prevent this, be sure all pipes are securely anchored along the run.

Water hammer may also be prevented or its severity reduced by installing an air chamber just ahead of the faucet. The air chamber may be a piece of air-filled pipe or tubing, about 2 feet long, extending vertically from the pipe. It must be airtight. Commercial devices designed to prevent water hammer, as discussed earlier, are also available.

Shrill noises like chattering or whistling may be traced to a faucet that has a loose washer. Other types of water noises, such as tapping, knocking, and rumbling, call for careful study of the situation to localize it and track down its cause—finding the cause will usually suggest the cure. One kind of rumbling noise has an ominous significance. This is caused by steam in the hot-water lines when the aquastat of a hot-water heater is not functioning properly. A faulty aquastat should be replaced immediately. While the pressure relief valve on the heater tank serves as a protection against undue hazard, the steam may rupture the pipes and cause other damage.

Gurgling of water drains may indicate that the drain trap under the sink is partially clogged, slowing the water. But, the more likely possibility is that there is a conflict of water trying to go out and air coming in. This may occur because the sink drain is not connected to the vent stack or because the vent pipe is plugged with insect nests or waste forced up from below while cleaning out a clogged drain.

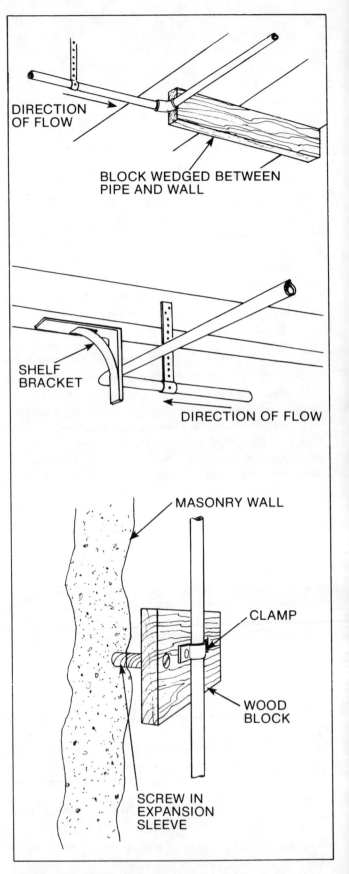

Methods of securing pipes. Persistent creaking noises are a sign of loose pipes.

Index